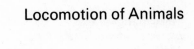

Locomotion of Animals

TERTIARY LEVEL BIOLOGY

A series covering selected areas of biology at advanced undergraduate level. While designed specifically for course options at this level within Universities and Polytechnics, the series will be of great value to specialists and research workers in other fields who require a knowledge of the essentials of a subject.

Titles in the series:

Experimentation in Biology	Ridgman
Methods in Experimental Biology	Ralph
Visceral Muscle	Huddart and Hunt
Biological Membranes	Harrison and Lunt
Comparative Immunobiology	Manning and Turner
Water and Plants	Meidner and Sheriff
Biology of Nematodes	Croll and Matthews
An Introduction to Biological Rhythms	Saunders
Biology of Ageing	Lamb
Biology of Reproduction	Hogarth
An Introduction to Marine Science	Meadows and Campbell
Biology of Fresh Waters	Maitland
An Introduction to Developmental Biology	Ede
Physiology of Parasites	Chappell
Neurosecretion	Maddrell and Nordmann
Biology of Communication	Lewis and Gower
Population Genetics	Gale
Structure and Biochemistry of Cell Organelles	Reid
Developmental Microbiology	Peberdy
Genetics of Microbes	Bainbridge
Biological Functions of Carbohydrates	Candy
Endocrinology	Goldsworthy, Robinson and Mordue
The Estuarine Ecosystem	McLusky
Animal Osmoregulation	Rankin and Davenport
Molecular Enzymology	Wharton and Eisenthal
Environmental Microbiology	Grant and Long
The Genetic Basis of Development	Stewart and Hunt

Locomotion of Animals

R. McNeill Alexander, D.Sc.

Professor of Zoology
University of Leeds

Blackie

Glasgow and London

Distributed in the USA
by Chapman and Hall
New York

Blackie & Son Limited
Bishopbriggs, Glasgow G64 2NZ
Furnival House, 14–18 High Holborn, London WC1V 6BX

Distributed in the USA by
Chapman and Hall, 733 Third Avenue,
New York, N.Y. 10017
in association with Methuen, Inc.

QP
301
, A296
1982

© 1982 Blackie & Son Limited
First published 1982

British Library Cataloguing in Publication Data

Alexander, R. McNeill
    Locomotion of animals.—(Tertiary level biology)
    1. Animal locomotion
    I. Title   II. Series
    591.1′852   QP301

ISBN 0-216-91159-1
ISBN 0-216-91158-3 Pbk

For the USA, International Standard Book Numbers are
cloth 0–412–00001–6
paper 0–412–00011–3

Filmset by Advanced Filmsetters (Glasgow) Ltd

Printed in Great Britain by
Thomson Litho Ltd, East Kilbride, Scotland

Preface

This book is about how animals travel around on land, in water and in the air. It is mainly about mechanisms of locomotion, their limitations and their energy requirements. There is some information about muscle physiology in Chapter 1, but only as much as seems necessary for the discussions of mechanisms and energetics. There is information in later chapters about the patterns of repetitive movement involved, for instance, in different gaits, but nothing about nervous mechanisms of coordination. I have tried to include most of the widely-used methods of locomotion, but have not thought it sensible to try to mention every variety of locomotion used by animals.

This book is designed for undergraduates, but I hope that other people will also find it interesting.

It is possible and sometimes illuminating to use complex mathematics in discussions of animal locomotion. This book includes many equations, but little mathematics. Such mathematics as there is, is simple.

Discussions of the mechanisms and energetics of locomotion inevitably involve mechanics. I expect that some readers will know a lot of mechanics, and some hardly any. I have tried to help the latter without boring the former, by putting a summary of the necessary mechanics in an appendix (p. 140). References from the main text to the appendix will tell readers where they can find help, if they need it. Figures and equations in the appendix have numbers distinguished by a prefix A (for instance, Figure A.3).

There is a list of references and suggestions for further reading at the end of the book. References in the text indicate principal sources of information, but I have not thought it desirable to insert the much larger number of references that would be needed to show the authority for every statement.

Dr H. C. Bennet-Clark and Dr C. J. Pennycuick read the typescript and made many very helpful suggestions.

R. McNeill Alexander

Contents

CHAPTER ONE

SOURCES OF POWER

The movements described in this book are powered by muscles, cilia, flagella or cytoplasmic streaming. This chapter is mainly about muscle but includes brief accounts of the other sources of power.

Muscle

Muscle fibres do work by shortening while exerting a force. The structures that generate force have been identified in electron microscope sections and are shown diagrammatically in Figure 1.1a,b (see also White, 1977). They are protein filaments a few micrometres long, arranged parallel to the length of the fibre. There are thick filaments of myosin and thin ones of actin and tropomyosin. The muscles that move the skeletons of vertebrates and insects are of the cross-striated kind shown in the figure. Interdigitating bands of thick and thin filaments alternate along the length of the fibre, and partitions called Z-discs cross the fibre, through the mid-points of the thin filaments. Each repeat of the resulting pattern of stripes is called a sarcomere. A different, obliquely-striated type of muscle fibre is found in nematode and annelid worms and in many of the muscles of molluscs. Its sarcomeres run obliquely across the fibre.

The thick filaments have projections called cross-bridges that seem to be capable of attaching to a thin filament and then swinging through an angle, making the filaments slide past each other. They can detach, swing back and reattach to a different point on the thin filament. In this way the cross-bridges can pull on the thin filaments like a man pulling in a rope hand over hand. Each sarcomere can be shortened from the state shown in Figure 1.1a to that shown in Figure 1.1b. Notice that if the sarcomere were stretched beyond the state shown in Figure 1.1a the thin filaments

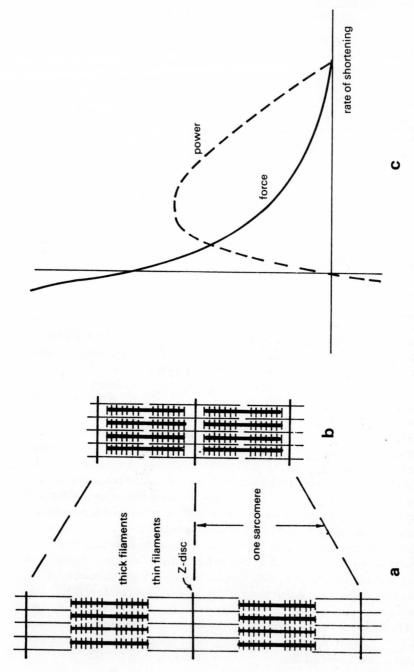

Figure 1.1 (a, b) Diagrams of a longitudinal section of cross-striated muscle. In (a) the muscle is extended and in (b) considerably shortened. (c) A schematic graph showing how the force and power that a muscle can exert depend on the rate at which it is shortening.

would be pulled right out from between the thick ones and the cross-bridges could no longer attach. Also, sarcomeres cannot shorten much beyond the point shown in Figure 1.1b because the thick filaments abut against the Z-discs. Therefore, cross-striated muscle can operate only over a limited range of lengths: the minimum length is not much less than half the maximum. There is an intermediate length at which all the cross-bridges can attach but none of the disadvantages of excessive shortening operate. This is the length at which most force can be exerted.

Since the cross-bridges transmit the force from the thick filaments to the thin ones, the force that a muscle fibre can produce is proportional to the number of cross-bridges in each sarcomere. Muscles with long filaments can therefore exert more stress than muscles with shorter ones. Vertebrate skeletal muscles have thick filaments about 1.6 μm long and can exert maximum stresses of about 0.3 MPa,* but locust leg muscle has filaments about 5.5 μm long and can exert 0.8 MPa.

Imagine (if you can) a vertebrate muscle of volume $1 m^3$. When this particular muscle is fully extended its fibres are 1.4 m long, and when it is fully shortened they are just half as long, or 0.7 m. About halfway through the contraction the fibres are 1 m long so the cross-sectional area is $1 m^2$, and if the stress is 0.3 MPa the force is 0.3 MN. The fibres shorten by 0.7 m while exerting this force, so the work done is $0.7 \times 0.3 = 0.21$ MJ. The mass of $1 m^3$ muscle would be about 1000 kg, so this work is 210 J per kg muscle. This is a very crude argument, but a more precise one leads to a similar conclusion. Vertebrate skeletal muscle cannot do more work than about 200 J/kg in a single contraction.

There is a limit to the rate at which cross-bridges can attach, pull and detach again. There is therefore a limit to the rate at which they can make the filaments slide past each other. The rate varies in different muscles: let it be u in a muscle fibre that has length l when its sarcomeres have length λ. Each sarcomere can shorten at a speed $2u$ and the fibre as a whole can shorten at $2ul/\lambda$ because it has l/λ sarcomeres. The fastest known muscles seem to be a rat eye muscle and a mouse toe muscle, which can shorten at rates corresponding to 25 fibre lengths per second (i.e. $2u/\lambda = 25 s^{-1}$). Mammal muscles have sarcomeres about 2.5 μm long, so this implies a sliding speed u of about 30 μm/s. Notice that the rate of shortening of the fibre would be less (for given u) if the sarcomere were longer. Long sarcomeres give high stresses but low rates of shortening.

The faster a muscle is shortening, the less force it can exert. Conversely,

---

* A megapascal (MPa) is one million newtons per square metre or 102 grams weight per square millimetre. See the explanation of units in the appendix.

the less resistance it experiences, the faster it can shorten (Figure 1.1c). The maximum stress of 0.3 MPa exerted by typical vertebrate skeletal muscles can be exerted only when the muscle is shortening very slowly or not at all. The maximum work of $200\,J\,kg^{-1}$ can therefore be done only in a very slow contraction. The maximum speed of shortening (25 lengths s^{-1}) given above for the fastest rat and mouse muscles can be achieved only when no external load acts on the muscle.

The power output of a muscle is the force it exerts multiplied by the rate of shortening. At the highest rate of shortening no force is developed, so the power is zero. A large force can be exerted if the rate of shortening is zero but the power is then again zero. Power output is greatest at an intermediate speed (Figure 1.1c). The flight muscles of various insects and hummingbirds have been shown to be capable of power outputs of about 200 W per kg muscle—this is the mean power output over a series of cycles of contraction and relaxation, and no muscles are known to produce much more power than this. These muscles work aerobically, and have up to about half of their volume occupied by mitochondria. It is possible that a very fast anaerobic muscle containing fewer mitochondria and more thick and thin filaments could produce more power. Aerobic and anaerobic muscles are discussed in the next section of this chapter.

A muscle that shortens while exerting a force does (positive) work. One that is forcibly extended while exerting a force has work done on it and degrades this work to heat (just as work is degraded to heat in the brakes of a vehicle). This is often described by saying that the muscle does *negative* work. Figure 1.1c shows that large forces cause lengthening (negative rates of shortening) of muscles. Muscles have to do negative work when a running animal slows down, or decelerates a moving limb.

The rates at which animals use chemical energy can be calculated from their rates of oxygen consumption (unless anaerobic metabolism is being used). This is the basis for many measurements of muscle efficiency. For instance, the rates of oxygen consumption have been measured for people resting and doing work at a known rate by pedalling a bicycle ergometer. The difference gave the metabolic power used for pedalling, so the efficiency could be calculated as (mechanical power output) $\div$ (metabolic power consumption). This experiment showed that over a wide range of rates of pedalling, the efficiency of doing positive work is about 0.25. Measurements of the rates of oxygen consumption of people, squirrels and other mammals running up steep gradients, and of birds flying up gradients in wind-tunnels, have confirmed that muscle can do positive work with efficiencies up to about 0.25. (Experiments involving running up shallow gradients give deceptively high efficiencies because such gradients reduce the negative work performed in each stride as well as increasing the positive work—see Margaria, 1976.)

It might be thought that an animal doing negative work would use less metabolic energy, and less oxygen, than when resting. This is not so, because metabolic energy is needed to maintain tension in the muscles. Since (positive) metabolic energy is used doing negative work, the efficiency of doing negative work is negative. Measurements of the rates of oxygen consumption of humans and squirrels running down steep slopes show that 1 J metabolism is sufficient for about 1.2 J negative work: muscles can do negative work with efficiencies around −1.2.

Types of muscle

This section is about peculiarities of different types of muscle. In aerobic metabolism, energy is released from foodstuffs by oxidizing them to carbon dioxide, water, etc. If an animal relies entirely on aerobic metabolism, the rate at which it can use energy for locomotion or any other purpose depends on the rate at which oxygen can be brought to the tissues. This in turn is limited by the rate at which the respiratory organs can take up oxygen, and the rate at which the blood can transport it round the body, in animals whose blood has this function. It is limited in insects by the rate at which oxygen can diffuse or be pumped through the tracheal system. Vertebrates and crabs have escaped from this limitation by evolving the ability to carry oxygen debts (Ruben and Bennett, 1980). In bursts of activity too violent to be supported by aerobic metabolism, they resort to an anaerobic process: they convert glucose to lactic acid, which releases energy without requiring oxygen. The energy released is much less than the heat of combustion of glucose but the residue is not lost. In a subsequent period of rest, much of the lactic acid is reconverted to glucose while the remainder is oxidized to carbon dioxide to supply the energy required. The oxygen used in this period of recovery can be thought of as repaying an oxygen debt. Only a limited amount of anaerobic metabolism is possible before a period of rest, because excessive concentrations of lactic acid cannot be tolerated. An athlete breathing heavily after a race is repaying an oxygen debt.

Vertebrate muscle fibres that work by anaerobic metabolism tend to be white and those that work aerobically tend to be red. The difference is easy to see when eating chicken or fish. The white meat of chicken breast (the principle flight muscles) is clearly different from the dark (red) meat of the legs. Most of the flesh of fish is white muscle but many species have a conspicuous line of red muscle immediately under the skin on each side of the body. It has been shown, by recording electrical activity from the muscles of dogfish (*Scyliorhinus*) and other fish, that only the red muscle is used in prolonged slow swimming (Bone, 1966). The much larger quantity of white muscle is brought into use for short bursts of faster swimming that cannot be sustained for long. The red muscle has a better blood supply than the white and contains far more mitochondria, the organelles that contain the enzymes of the Krebs cycle which are involved in aerobic metabolism.

Muscle fibres also differ in the speeds at which they can shorten. Small animals tend to make repetitive movements at higher frequencies than large ones. For instance, running mice make up to 7.5 strides per second but adult elephants (*Loxodonta*) seem not to use frequencies above 1.4 Hz (cycles per second). Budgerigars (*Melopsittacus*) beat their wings at about 14 Hz but condors (*Vultur*) beat theirs at only 2.5 Hz. Correspondingly, small animals tend to have faster muscles than large ones. The fibres of the soleus muscles of mice, rats and cats shorten at maximum speeds of about 11, 7 and 5 lengths per second respectively.

There are differences of speed between muscle fibres in the same animal, as well as between animals. Mammals have three types of fibre in their skeletal muscles. The SO (slow oxidative) fibres are relatively slow and depend on aerobic metabolism. They are the predominant fibre type in the soleus muscle. The FOG (fast oxidative glycolytic) fibres are faster and can work either aerobically or anaerobically. Both SO and FOG fibres are red and contain plenty of mitochondria. Finally, FG (fast glycolytic) fibres are also fast but work mainly anaerobically. They contain few mitochondria. The extensor digitorum longus (a toe muscle) includes a large proportion of FG fibres and can shorten about twice as fast as the soleus in mice and rats. All three fibre types are found in the leg muscles of mammals, in different proportions in different muscles. Experiments with rats have shown that most of the SO and FOG fibres of the ankle extensor muscles are used at all speeds of running. Most of the FG fibres are used in fast running but only a few in slow running (Armstrong, 1981).

The wing muscles of locusts and other relatively primitive insects have properties very like vertebrate skeletal muscle, but those of flies (Diptera) and other more advanced insects have remarkable properties which enable them to work at very high frequencies, for instance 120 Hz for a blowfly (*Sarcophaga*) and 600 Hz for a mosquito (*Aedes*). These are called fibrillar muscles. Successive contractions of fibrillar muscle fibres do not have to be fired by separate nerve action potentials (Usherwood, 1975)—a blowfly muscle working at 120 Hz may be receiving action potentials at only 3 Hz. These occasional action potentials suffice to keep the muscle in a state that makes oscillatory contraction possible. Any system that has mass and elastic properties has a natural frequency of oscillation. For instance, a mass hanging on a spring will bounce up and down at its natural frequency if it is once set into motion. When fibrillar flight muscle is active, it contracts and relaxes at the natural frequency of any system of which it is part. It maintains the oscillations instead of letting them die away as the oscillations of a mass on a spring would do.

Arrangement of muscles

Vertebrates and arthropods have skeletons consisting of stiff elements connected by flexible joints (see Currey, 1980). The simplest common type of joint is a hinge, which allows only rotation about a single axis. It is possible for a single muscle to work a hinge joint, if the joint is spring-loaded. The joints between the two valves of the shells of bivalve molluscs are like this (Figure 1.2a). Contraction of the adductor muscle closes the shell, compressing a block of a rubber-like protein called abductin. When the muscle relaxes the shell is opened by the elastic recoil of the abductin.

Most hinge joints are not spring-loaded and need at least two muscles, a flexor to bend and an extensor to extend them (Figure 1.2b). Two muscles with opposite effects like this are called antagonists. Joints allowing a wider variety of movements need more muscles. The human knee is a hinge allowing rotation about one axis, the wrist is a universal joint allowing rotation about two axes and the hip is a ball-and-socket joint allowing rotation about any axis. Some other joints allow sliding movements as well as rotations.

Figure 1.2c shows a muscle that has a moment arm r about a joint. (This is the perpendicular distance from the line of action of the muscle to the axis of the joint.) The volume of the muscle is V and the present length of its fibres is l, so its cross-sectional area is V/l. If the stress in the muscle is σ the force is $V\sigma/l$ and the moment about the joint is $V\sigma r/l$. If the muscle shortens by a fraction ε of its present length, that is by a distance εl, it rotates the joint through an angle of about $\varepsilon l/r$ (measured in radians). Neither the moment nor the angle is changed by changing the initial fibre length l, provided the moment arm r is changed so as to keep l/r constant. A muscle with long fibres and a large moment arm may be precisely equivalent to one with short fibres and a small moment arm. The pennate

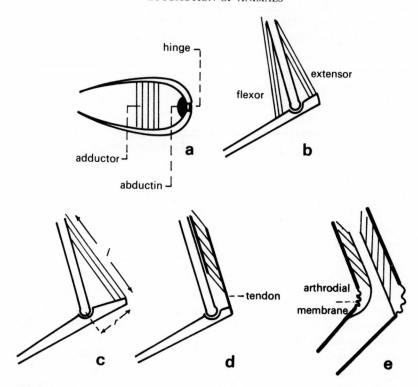

Figure 1.2   Diagrams of (a) the shell and adductor muscle of a bivalve mollusc, in section; (b) a hinge joint in a vertebrate limb, with its muscles; (c) a parallel-fibred muscle that extends a joint; (d) an equivalent pennate muscle; (e) a hinge joint in an arthropod limb, and its muscles, in section.

muscle shown in Figure 1.2d may be equivalent to the parallel-fibred one shown in Figure 1.2c. Pennate muscles have their fibres attaching obliquely to tendons, which are often long. The elastic properties of the tendons are sometimes useful, for instance in mammal running and locust jumping (Chapter 5).

Crabs, insects and other arthropods have slender legs with tubular skeletons enclosing the muscles (Figure 1.2e). The tubular segments are hinged together at the joints, where flexible arthrodial membranes fill the gaps between them. The muscles cannot be stout and cannot have large moment arms, because the tubes are slender. Pennate muscles fit conveniently into the legs, and most arthropod leg muscles are pennate.

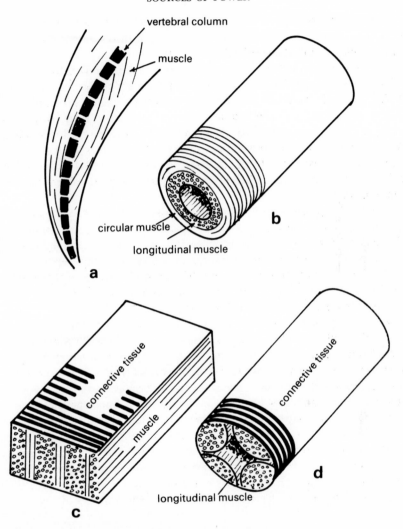

Figure 1.3 Diagrams showing antagonistic groups of muscle fibres in (a) the tail of a fish (seen in horizontal section); (b) a worm-like animal; (c) a block of tissue; and (d) a different worm-like animal. These diagrams are explained further in the text.

The backbones of vertebrates allow a little movement between each vertebra and the next. Contraction of the muscles of the right side bends the backbone to the right (Figure 1.3a) and the muscles of the left side bend it to the left. Thus the left and right muscles are antagonistic.

The diagram indicates that the muscle fibres are not all parallel to the backbone, but cannot give a clear impression of their complicated three-dimensional arrangement (Alexander, 1969).

Muscles can be antagonistic even without a jointed skeleton (Clark, 1964). Most of the materials of which animals are made are liquids or solids, so their volumes are scarcely altered by changes of pressure. Many worms have longitudinal and circular muscles in their body walls (Figure 1.3b). They can shorten themselves by contracting the longitudinal muscles but this makes them fatter, because their volumes are fixed. They can make themselves thinner by contracting the circular muscles but this also lengthens them. Thus the circular and longitudinal muscles are antagonistic.

A different application of the same principle is illustrated by Figure 1.3c. The block of tissue has its width kept constant, for instance by connective tissue fibres running transversely across it. Contraction of the longitudinal muscles makes it shorter but thicker, and contraction of the vertical ones makes it longer but thinner. The walls of the mantle cavities of cephalopod molluscs and the feet of many other molluscs have muscle fibre arrangements of this general type (Ward, 1972).

Figure 1.3d shows how bending is possible for animals without skeletons. The animal illustrated has longitudinal muscles and cuticle with inextensible circular fibres. If the muscles of the left side contract they bend the animal to the left and if those on the right contract they bend it to the right. The muscles of the two sides are antagonistic because the volume is fixed and the cuticle prevents any increase in diameter. Shortening of the left side must be accompanied by lengthening of the right side. This kind of arrangement is found in nematode worms which, however, have more complex cuticle structure. Also, these worms bend dorsally and ventrally rather than from side to side, in their crawling and swimming movements. The same principle can be used by other worms with extensible cuticles, if they have circular muscles and use them to prevent increases of diameter.

Other sources of power

Many small animals swim by means of flagella or cilia, which are long slender structures of diameter about 0.25 μm. Examples of animals with flagella and cilia are shown in Figures 2.3 and 2.11 respectively. Flagella make undulating (eel-like) movements and cilia make oar-like movements, which propel the animals (Holberton, 1977). Figure 1.4a shows part of a flagellum or cilium, enormously magnified. Running lengthwise along it are microtubules, hollow tubes of diameter about 20 nm. There are nine double microtubules around the circumference and two single ones in the centre. Each double microtubule is fringed along one side by projections called dynein arms.

It has been shown by studying electron micrographs of bent cilia that bending involves no change in the lengths of microtubules. Rather, the microtubules slide relative to each other. Figure 1.4b shows that if the microtubules on the right slide down towards the base, the cilium must bend to the right. It is believed that the sliding is caused by the action of dynein arms on adjacent microtubules, just as sliding in muscle is caused by cross-bridges acting on thin filaments. Some flagella beat at very high frequencies, around 70 Hz, and it can be calculated that in them the speed of sliding of adjacent microtubules relative to each other reaches 20 μm/s. This is similar to the maximum speeds of sliding of thin filaments relative to thick ones in fast muscles.

Amoebas crawl by cytoplasmic streaming, that is by flow of cytoplasm within the moving cell (Holberton, 1977). The mechanism is not understood but seems to have a good deal in common with muscular con-

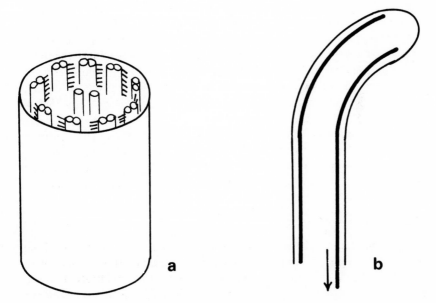

Figure 1.4   (a) Diagram of part of a flagellum or cilium.
(b) Diagram showing how bending is caused by sliding microtubules.

traction. Electron micrographs of amoebas show thick and thin filaments scattered in the cytoplasm, with diameters similar to the thick and thin filaments of muscle. The thin filaments seem to be chemically similar to those of muscle. It has been shown that the protein of muscle cross-bridges will attach to them, just as it attaches to the thin filaments of muscle. There are indications that the filaments may not be permanent, but may dissociate into component molecules and re-form in the course of crawling.

CHAPTER TWO

SWIMMING

Animals swim by undulation, by rowing, by means of hydrofoils and by jet propulsion. These four main mechanisms of swimming will be discussed in turn in this chapter; a separate chapter on buoyancy follows.

Swimming by undulation

Figure 2.1 shows a spermatozoon and an eel-like fish swimming. Both are long and slender, and both swim by wave-like movements which travel backwards along the body, pushing the water backwards and the organism forwards. However, in two important respects the sperm and the fish are very different. First, the tail of the sperm is a flagellum which works as described on p. 10, but the tail of the fish is moved by muscle. Secondly, the sperm is small and slow but the fish is relatively large and fast. (The spermatozoon may seem to move quite fast under the microscope which magnifies the distance it travels, but its speed is actually only 0.1 mm s^{-1}.) This difference may seem trivial but it is so fundamental that large and small undulating organisms will have to be discussed separately.

Differences like this are best expressed by calculating a quantity called the Reynolds number, which depends on the size and speed of the body, as explained in the appendix, p. 148. The pattern of flow of water around similarly shaped animals, swimming by similar movements, is the same only if the Reynolds numbers for the movements are equal. Small Reynolds numbers (< 1) imply that the water movements are controlled principally by viscosity and large ones ($\gg 1$) that inertia is more important. A sperm 60 µm long swimming at 0.1 mm/s has a Reynolds number of 0.006. A small fish 150 mm long swimming at 120 mm/s has a Reynolds number of 18 000. Consequently, the inertia of the organism and the

13

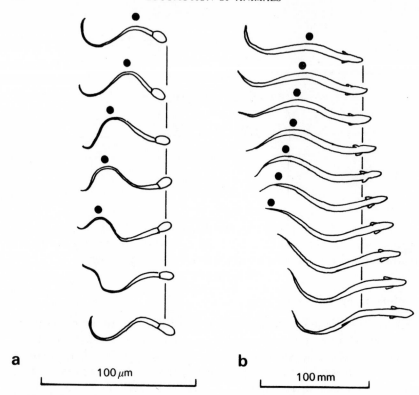

a   **b**

| 100 μm |

| 100 mm |

Figure 2.1 Outlines traced from cine films of two organisms swimming by undulation. In both cases the wave crests (marked by dots) travel backwards along the body.
(a) A bull spermatozoon. Interval between pictures 15 ms.
(b) A butterfish, *Centronotus*. Interval 50 ms.
Redrawn from (a) J. Gray (1958) *J. exp. Biol.* **35**, 96–108 and (b) J. Gray (1933) *J. exp. Biol.* **10**, 88–104.

surrounding water can be neglected in discussions of sperm swimming, but not of eels swimming. This would be illustrated rather dramatically if both organisms suddenly stopped undulating: the sperm would stop almost instantaneously but the eel would glide to a halt.

The Reynolds numbers quoted in the previous paragraph were calculated from the length of the organism and its forward speed. Since swimming depends on the side-to-side movement of the tail, it might have been more appropriate to have used the width of the tail and the speed of its side-to-side movements. Smaller Reynolds numbers would have been obtained in each case but the general conclusion would have been the

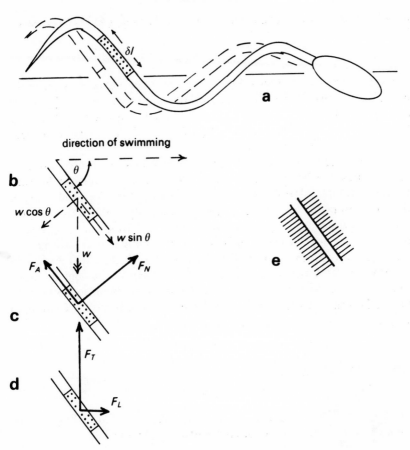

Figure 2.2 (a) to (d) Diagrams of a spermatozoon swimming. In (a), two successive positions are represented by the continuous and broken outlines, in that order: (b) shows the components of the velocity of a segment and (c), (d) show the hydrodynamic force on it resolved into components in two different ways; (e) shows part of a flagellum with flimmer filaments.

same: the Reynolds number is much less than 1 for the sperm but much more than 1 for the fish, so inertia can be ignored in the first case but not in the second.

Flagellates and sperm

Consider the sperm first. Since the Reynolds number is so low the forces on the tail can be calculated from equation A.14 (see Appendix): the drag on a body of length l moving with velocity u through a fluid of viscosity η is given by

$$\text{drag} = k\eta l u \qquad 2.1$$

The value of the constant k depends on the shape and orientation of the body. Let the value for a cylindrical rod moving lengthwise along its own axis be k_A (A stands for axial). Let the value for the same rod moving broadside on, normal to its axis be k_N (N stands for normal). It has been shown theoretically that $k_N \simeq 2k_A$.

In Figure 2.2a, a spermatozoon is passing waves backwards along its tail, towards the left of the diagram. How does this move it forward towards the right? We will suppose it is initially stationary and discover the direction of the force generated by its undulating movements. At the instant illustrated, the stippled segment of the tail (which has length δl) is inclined at an angle θ to the direction of swimming and is moving with velocity w towards the bottom of the page (Figure 2.2b). This velocity can be resolved into a component $w \sin \theta$ along the axis of the segment and a component $w \cos \theta$ at right angles to it. The corresponding components of the drag on the segment can be calculated from equation 2.1. There is an axial component F_A (see Figure 2.2c) given by

$$F_A = k_A \eta w \sin \theta \cdot \delta l \qquad 2.2$$

and a normal component

$$F_N = k_N \eta w \cos \theta \cdot \delta l \qquad 2.3$$

The force on the segment can alternatively be resolved into a longitudinal component F_L and a transverse component F_T (Figure 2.2d).

$$\begin{aligned} F_L &= F_N \sin \theta - F_A \cos \theta \\ &= (k_N - k_A)\eta w \sin \theta \cos \theta \cdot \delta l \end{aligned} \qquad 2.4$$

Since k_N is larger than k_A, F_L is positive: the movements of the sperm tend to propel it forwards. Since some parts of the body, at any instant, are moving to the sperm's left and some to its right, transverse forces F_T on different parts of the body tend to cancel each other out, and the effect of the swimming movements is simply to drive the organism forwards.

Figure 2.3 compares a spermatozoon with some other organisms that swim by undulating flagella. All operate at low Reynolds numbers. The sperm swims head first with the flagellum (tail) pushing it along from behind. *Strigomonas* swims with its flagellum in front, pulling it along. *Polytoma* has two flagella, one on each side. In all these cases

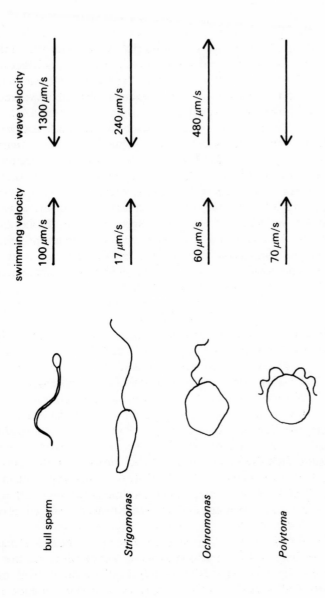

Figure 2.3 Four organisms that swim by undulating flagella. The direction of swimming, and the direction of movement of waves along the flagella, are indicated by arrows.

the animal moves in the opposite direction to the waves, as expected from equation 2.4.

Ochromonas (also shown in Figure 2.3) is different: it travels in the same direction as the waves on its long flagellum. How can this be? The argument leading up to equation 2.4 shows that organisms should travel in the opposite direction to their flagellar waves provided k_N is larger than k_A. For smooth rods, $k_N \simeq 2k_A$. The short flagellum of *Ochromonas* is smooth but it seems to play no part in swimming. The long (swimming) flagellum is not smooth, but has numerous hair-like projections which are called flimmer filaments (Figure 2.2e) These are so thin that they can be detected only by electron microscopy, and their position during swimming cannot be seen in photographs. A likely explanation of the direction of swimming depends on the assumption that they lie in the plane of beating of the flagellum (i.e. in the plane of the paper in Figure 2.3).

The flimmer filaments are only about 1 μm long but there are so many of them that they are in aggregate about 20 times as long as the flagellum. Consequently the hydrodynamic forces on them must be much larger than the forces on the main strand of the flagellum. For a naked flagellum $k_N \simeq 2k_A$. For one bearing numerous flimmer filaments at right angles to the main strand, $k_A \simeq 2k_N$ and equation 2.4 predicts a negative propulsive force. In other words, it predicts that the animal will be propelled in the direction of the waves, as observed for *Ochromonas*.

How much power does swimming with a flagellum need? Have flagellates evolved to be as fast and as economical of energy as possible? Power is (force × speed) so the power required to propel a body through water is (drag × speed) which is $k\eta l u^2$ if the Reynolds number is low (from equation 2.1). For a rod moving lengthwise the appropriate value of k is k_A, and for rods with the proportions of a flagellum (with the length a few hundred times the diameter) $k_A \simeq 1$. If a flagellum was simply a rigid rod moving lengthwise through the water, the power needed to propel it would be about $\eta l u^2$. However, the mechanism of propulsion requires the flagellum to move from side to side. The side-to-side movements are much faster than the forward speed (Figure 2.1a) so the power required is far more than $\eta l u^2$: it has been calculated that it is approximately $50\eta l u^2$ (Wu, 1977).

Since the head of a sperm is so small, the power needed to propel a sperm is little different from the power which would be needed for an isolated, swimming flagellum. In contrast, flagellate protozoans have relatively large bodies, many of them roughly spherical. The constant k for spheres is about 9, so the power needed to propel a body of diameter d is about $9\eta du^2$. The power needed to propel a flagellate can be estimated by adding this to the power $50\eta lu^2$ needed to propel the flagellum alone. Most flagellates have flagella at least as long as their bodies, i.e. $l \geqslant d$. Consequently the need to propel the body adds 18% or less to the power which would be needed to propel the flagellum alone.

All flagella (except those of bacteria) have essentially the same structure (Figure 1.4). It seems reasonable to guess that the power they can produce is proportional to their length. Most of the mechanical power needed for swimming is the power needed to propel the flagellum itself, $50\eta lu^2$. This is also proportional to the length of the flagellum, so all sperm and flagellates of whatever size might be expected to have about the same maximum speed. Most observed speeds lie between 50 and 150 μm/s, although some flagellates such as *Strigomonas* (Figure 2.3) are slower. There is no obvious correlation between size and speed.

Since most of the power is needed to propel the flagellum itself, a flagellate could increase its speed very little by evolving more or longer flagella. On the other hand, it could seemingly reduce its energy consumption without much loss of speed by evolving shorter flagella. Why has this not happened? The answer seems to be that the energy required for swimming is only a trivial fraction of the total metabolism of the organism. A structure as small as a flagellum, which represents only a tiny fraction of the volume of the organism, cannot consume much power.

Fish and worms

Figure 2.1 showed that the swimming movements of sperm and eel-like fish are very similar. Similar movements are also used for swimming by water snakes, leeches and some other sorts of worms, though leeches bend their bodies dorso-ventrally in contrast to eels and snakes, which bend from side to side. The Reynolds numbers of swimming eels, snakes and all but the smallest worms are too high for equations 2.2 and 2.3 to apply to them. Nevertheless, part of the explanation of swimming by undulating flagella applies also to these animals: waves travelling backwards along their bodies propel them forward because, even at high Reynolds numbers, a rod moving lengthwise experiences less drag than one moving broadside-on at the same speed.

Many polychaete worms have large flap-like parapodia projecting from their sides (Figure 2.4). Some of them swim by undulating their bodies from side to side but, unlike eels, are propelled forward by waves travelling forward along their bodies (Clark and Tritton, 1970). This is partly

Figure 2.4 A polychaete worm, *Nereis diversicolor*, swimming towards the right of the page. This sketch is based on a photograph published by R. B. Clark and D. J. Tritton (1970) *J. Zool., Lond.* **161**, 257–271.

because the parapodia act like the flimmer filaments of *Ochromonas* (Figure 2.3), but the parapodia are also muscular and make rowing movements which supplement this effect.

The anterior parts of a swimming eel bend less than the posterior parts, so the waves increase in amplitude as they travel back along the body. This effect is so marked in more typical fish that it is not very obvious that the anterior parts are undulating at all (Figure 2.5b).

The forces on undulating flagella depend far more on viscosity than on inertia, because the Reynolds numbers are so low. For this reason, the equations which were used (2.1 to 2.4) included the viscosity but not the density of the water. In contrast, the movements of fish are to be explained mainly in terms of inertia (Webb, 1975; Hoar and Randall, 1978).

A swimming fish sets the water behind it moving. The water movements have been studied by the following method (McCutchen, 1977). The fish is put in an aquarium containing water which is warmer near the surface than deeper down. When it swims, it disturbs the water, and so warm and cool water are brought into contact. The refractive index of water depends on its temperature, so the places where warm and cold water meet can be made visible by suitable lighting. (Similarly, the surface of a pane of glass is visible because air and glass have different refractive indices.) Figure 2.5 is drawn from photographs obtained in this way. In Figure 2.5a the fish was swimming in intermittent fashion, making a couple of beats of its tail and then coasting along for a while before beating its tail again. Just before the photograph was taken it was coasting slowly towards the top of the page. It then made two beats of its tail, to left and then right, which set masses of water moving to either side, and also accelerated the fish and turned it towards the left. (The moving water presumably formed into vortex rings, but this is not obvious in a photograph—see p. 153.)

The Principle of Conservation of Momentum says that if no external

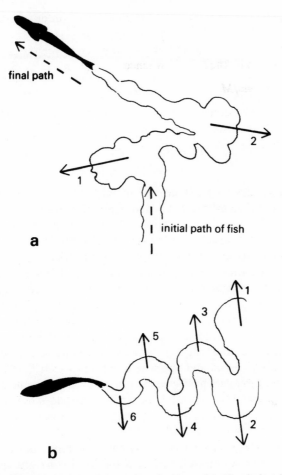

Figure 2.5 Outlines traced from photographs of a 32 mm fish (*Brachydanio*) swimming in apparatus which makes the moving water behind the fish visible. Arrows show the directions of water movement. The masses of water set moving by successive tail beats are numbered in order. From photographs by C. W. McCutchen (1977) in T. J. Pedley (edit.) *Scale Effects in Animal Locomotion*, Academic Press, London, pp. 339–363.

forces act on a system, its momentum remains constant. A fish can only give itself momentum by giving equal and opposite momentum to the water, and so to accelerate itself to the left it must accelerate water to the right as it did with its second tail beat (Figure 2.5a).

An important principle of swimming will be explained by taking a simpler example. Suppose a fish of mass m is initially stationary. It accelerates itself to velocity u by driving a mass M of water backwards with velocity $-U$. The momentum $-MU$ given to the water is equal and opposite to the momentum mu gained by the fish, whence

$$U = mu/M \qquad\qquad 2.5$$

The fish gives itself kinetic energy $\frac{1}{2}mu^2$, and gives kinetic energy $\frac{1}{2}MU^2$ to the water. This requires work W, where

$$W = \frac{1}{2}mu^2 + \frac{1}{2}MU^2$$
$$= \frac{1}{2}mu^2 [1 + (m/M)] \qquad\qquad 2.6$$

(using equation 2.5). A given amount of work will accelerate the fish to a higher velocity if the fish pushes a large mass M of water, than if it pushes a smaller mass. Consequently a large tail is a valuable aid to acceleration. It was found in a series of experiments that 18-cm trout (*Salmo gairdneri*) could accelerate from rest to a mean speed of 1.33 m/s, by one double tail beat (as in Figure 2.5a—Hoar and Randall, 1978). Similar fish with the anal fin and much of the tail fin removed could accelerate only to 1.07 m/s.

Figure 2.5b shows a fish swimming with approximately constant velocity. Since its momentum is constant, the total momentum of the water also remains constant. The leftward momentum given to the water by one tail beat is equal and opposite to the rightward momentum given by the next beat. Notice that the water is moving sideways, not backwards. Once it has reached the required velocity, the fish does not need to drive water backwards to maintain its velocity. It merely has to avoid slowing itself down by dragging water forwards.

The speed of the water moving sideways behind the fish could be measured from films obtained by the method of Figure 2.5. Alternatively it can be calculated from the movements of the tail seen in ordinary films (Lighthill, 1969; Wardle and Videler, 1980). The mass of water set moving by the tail in unit time can also be estimated from films: it is approximately the mass of water which passes through the imaginary hoop drawn in Figure 2.6, in unit time. This hoop is a circle of diameter equal to the span s of the tail, so its area is $\pi s^2/4$. If the fish is swimming at speed u through water of density ρ, the mass of water which passes through the hoop in unit time is $\pi \rho u s^2/4$. If this water is given velocity w (sideways) it is being given momentum at a rate $\pi \rho u w s^2/4$. By Newton's Second Law of Motion (equation A.7), the transverse force on the tail equals this rate of change of momentum. The power output P of the tail is the transverse force on it

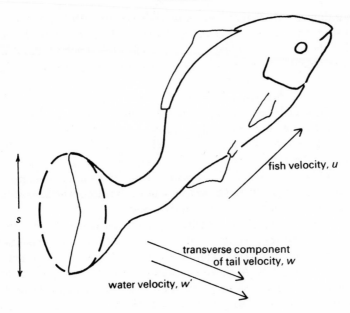

fish velocity, u

transverse component
of tail velocity, w

s

water velocity, w'

Figure 2.6 Diagram of a fish swimming (for details see the text).

multiplied by the transverse component w of its velocity,

$$P = \pi \rho u w w' s^2 / 4 \qquad 2.7$$

Many experiments have been performed on the swimming of trout 0.28 m long (Webb, 1971). These fish can swim indefinitely, without building up an oxygen debt, at speeds up to 0.6 m/s. They can reach 2.8 m/s in a short burst of speed lasting about one second. It can be calculated from films and equation 2.7 that their mechanical power output is 0.15 W at 0.6 m/s and 6 W at 2.8 m/s. Notice that the power which can be attained anaerobically at top speed is very much greater than the power which can be sustained by aerobic respiration.

It might seem reasonable to calculate the power needed for swimming in another, simpler way. Trout and many other fish are streamlined: they are roughly torpedo-shaped. The power needed to propel them might be expected to be about the same as for a rigid torpedo of the same size, which is little more than the product of the speed and the friction drag given by equation A.16. It is found, however, that the power calculated in this way for any particular speed is only about one-seventh of the power calculated from equation 2.7. Much more power is needed to propel a fish than would be needed to drive a torpedo of the same size at the same speed. It seems likely that the undulating movements of the fish thin the boundary layer and so increase the drag, but the theory of this has not been worked out in detail.

The fish shown in Figure 2.5a was alternately accelerating by making a few tail beats, and then coasting along without beating its tail. This style of swimming is often used by teleosts. It may save energy, because the drag probably falls to about the value expected for a rigid body while the fish is coasting (Wardle and Videler, 1980). This possibility has been checked by analysing a film of a cod (*Gadus morhua*). The deceleration during periods of coasting was measured and used to calculate the drag, which was found to have about the value expected for a rigid body.

In the range of speeds in which the power all comes from aerobic metabolism, the power used in swimming can be estimated from measurements of oxygen consumption. Apparatus designed for such measurements is shown in Figure 2.7a. A pump circulates water through the system of pipes which is designed to give smooth, even flow through the section containing the fish. The fish swims against the current so as to remain stationary relative to the apparatus. The grid behind it can be electrified if necessary to encourage it to keep swimming: if it allows itself to drift back against the grid it receives a mild electric shock. The oxygen electrode is used to monitor the decline in the dissolved oxygen content of the water due to the fish's respiration. Obviously the water must be replaced with fresh, aerated water before its oxygen content falls too low.

Figure 2.7b shows rates of oxygen consumption measured in this way for 0.28 m trout. The difference between the rates of oxygen consumption during swimming and at rest can be used to estimate the mechanical power output involved in swimming. This requires knowledge of the efficiency of the muscles, which has been measured by an ingenious experiment with plates attached to the backs of the trout (Webb, 1971; Figure 2.7c). These plates increased the drag on the fish by easily-calculated amounts. Consequently, oxygen was used faster at any particular

---

Figure 2.7   (a) Apparatus used for measuring the rates of oxygen consumption of swimming fish.

(b) Graph of rate of oxygen consumption against swimming speed for 0.28 m trout, *Salmo gairdneri*. (From P. W. Webb (1971) *J. exp. Biol.* **55**, 521–540.)

(c) Fish with a plate attached to its back, used in an experiment to measure the efficiency of swimming muscles.

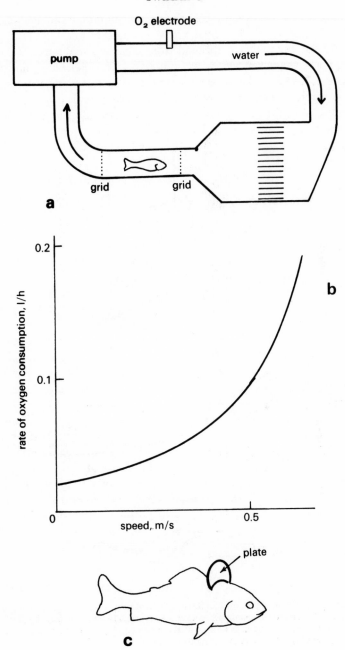

swimming speed. The increased oxygen consumption was measured so that the efficiency could be calculated. Mechanical power outputs in swimming without the plates, calculated from efficiency and rates of oxygen consumption, agree well with power outputs calculated from equation 2.7.

Figure 2.7b shows that a fish swimming quite slowly may use several times as much oxygen (and therefore energy) as when it is at rest. This is strikingly unlike flagellate protozoans which use only small fractions of their total energy consumption for swimming.

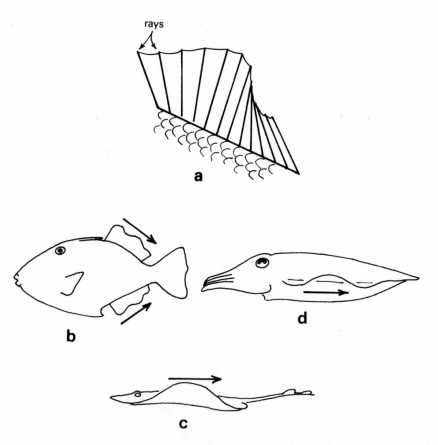

Figure 2.8   (a) Diagram of part of an undulating teleost fin.

   (b), (c), (d) Sketches of three animals swimming towards the left of the page by undulating fins. Arrows show the directions in which waves are moving along the fins. (b) is a trigger fish (*Balistidae*), (c) a ray (*Raia*) and (d) a cuttlefish (*Sepia*).

Many fish can propel themselves by undulating fins, as an alternative to undulating the whole body (Blake, 1978). The fins of teleost fish are fan-like structures consisting of webs of skin supported by widely-spaced bony rays (Figure 2.8a). Each ray has its own muscles and is joined to the skeleton of the trunk by a universal joint so that it can be swung forward and back or from side to side. The fin can be thrown into waves which can be made to travel along it. Many fish use this as a means of swimming. The trigger fish shown in Fig. 2.8b swims slowly by passing waves backwards along its pectoral, dorsal and anal fins. At higher speeds it folds the pectoral fins flat against the body and uses tail movements to supplement the action of the dorsal and anal fins. Rays have huge pectoral fins which are much less flexible than the fins of teleosts but are nevertheless capable of undulation (Figure 2.8c). They are the sole means of swimming, as the tail and its muscles are greatly reduced. Fin undulation is also used for swimming by cuttlefish (Figure 2.8d) which are molluscs, not fish. Cuttlefish also use jet propulsion: they swim slowly by fin undulation and fast by jet propulsion.

Cuttlefish and many teleosts have almost the same density as the water they swim in. For such animals, fin undulation provides a mechanism for subtle and precise manoeuvring. Trigger fish, for instance, swim backwards by sending waves forward along their dorsal and anal fins. They tilt their bodies in preparation for swimming up or down by sending waves forwards along one of these two fins and backwards along the other. They can pivot on a vertical axis through the body by appropriate movements of the dorsal, anal and one pectoral fin. These and other movements are all possible because fins are attached to various positions on the body and can each exert forces in various directions. Trigger fish live on coral reefs, feeding on algae and invertebrates, and the ability to manoeuvre precisely must be very useful to them as they hunt for food among the branches of coral.

Rowing

Figure 2.9 shows two organisms which propel themselves by rowing. *Chlamydomonas* is about 10 μm long and swims at about 40 μm/s, with a Reynolds number (based on body length) of about 4×10^{-4}. *Acilius* is about 17 mm long and swims at 50–500 mm/s, with Reynolds numbers of 900–9000. These Reynolds numbers imply that inertia is unimportant in the swimming of *Chlamydomonas*, but very important indeed in the swimming of *Acilius*.

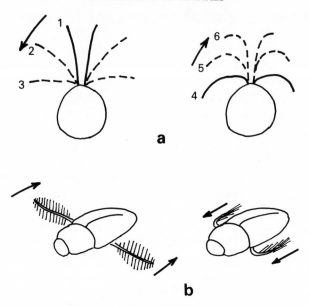

Figure 2.9 Stages in the swimming movements of (a) *Chlamydomonas*, a flagellate protozoan and (b) *Acilius*, a water beetle. Separate diagrams show the power stroke and the recovery stroke in each case. *Chlamydomonas* has just two flagella, which are shown in several successive positions (numbered) for each stroke.

Chlamydomonas has two flagella like *Polytoma* (Figure 2.3), but instead of undulating them it uses them like oars. In the power stroke they start at position 1 and are kept fairly straight except at the base, which bends posteriorly. In the recovery stroke a bend travels up each flagellum from base to tip, restoring it to its initial position. In the power stroke the flagellum is more or less at right angles to its path through the water. In the recovery stroke the part of the flagellum distal to the bend is more or less in line with its path through the water. When *Chlamydomonas* starts from rest the flagella meet more resistance in the power stroke than in the recovery stroke because $k_N > k_A$ (p. 16). The organism is therefore propelled forwards (towards the top of the page).

Acilius has three pairs of legs, like other insects, but only the hindmost pair (which is the most important in swimming) is shown in Figure 2.9b. These legs are rather flattened and are edged with hair-like setae which are hinged at the base in such a way that they spread during the power stroke but trail behind during the recovery stroke. Also, the flattened segments of

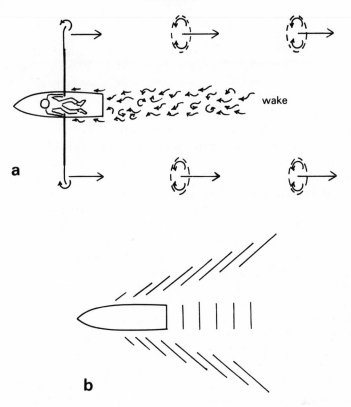

wake

a

b

Figure 2.10   Water movements due to the passage of a rowing boat. (a) shows movements within the water and (b) shows surface waves.

the legs turn so that they move broadside-on through the water in the power stroke but edge-on in the recovery stroke (Nachtigall, 1980). These movements of the legs and hairs make the force exerted on the water as large as possible in the power stroke but as small as possible in the recovery stroke.

Figure 2.10 shows water movements produced by a rowing boat. The surface waves do not concern us here because, unlike a boat, a water beetle swimming well below the surface produces no waves. Boats (and water beetles) leave wakes: water in the boundary layer is dragged along by the hull (or body) so that a wake of forward-moving water is left behind. At each stroke the oars (or legs) drive masses of water backwards. To keep the speed constant, the oars must give backward momentum to the water at the same rate as the hull transmits forward momentum to the wake. As in the case of the fish tail (p. 22), less power is needed if the oars accelerate large masses of water to a low speed than if they accelerate small masses to a high speed. For this reason, large oar blades are more efficient than small ones. The hind legs of water beetles, extended by their fringes of setae, make rather large oar blades.

All the kinetic energy left behind in the wake must be supplied by the work of rowing, so it is advantageous to reduce disturbance in the wake to the minimum possible. This can only be done by streamlining the hull. Wind-tunnel tests on the bodies of water beetles (with the legs removed) show that they are fairly well streamlined (Nachtigall, 1980). The drag on an *Acilius* body is about 2.5 times the drag on a well-streamlined body of the same cross-sectional area, at the same speed, but it is only about one-quarter as much as the drag on a sphere. The value of the stream-lining would be even more apparent if the beetle were compared with a sphere of the same volume, rather than with one of the same cross-sectional area. *Chlamydomonas* is almost spherical but a streamlined shape would give it no advantage because it swims at such low Reynolds numbers that there is no tendency for eddies to form in the wake.

An oarsman puts only the blades of his oars in the water, but a water beetle also has the shafts of its oars submerged. The blade of an oar and the outer part of the shaft move backward relative to the water in the power stroke and forward in the recovery stroke, but the inner part of the shaft moves at almost the speed of the hull and so travels forward all the time. It cannot therefore contribute to the propulsive force and is best made as slender as strength permits. Appropriately, water beetles have no setae on the proximal segments of their legs.

Some fish, including the angel fish *Pterophyllum*, swim slowly by rowing themselves along with their pectoral fins. The mechanical principles are the same as for the rowing of water beetles.

Ciliated animals

Ciliate protozoans such as *Paramecium* (Figure 2.11a) have their bodies covered by cilia, which are typically $10\,\mu m$ long and about $2\,\mu m$ apart. They have the typical $9+2$ structure of flagella (Figure 1.4) and they beat in essentially the same way as the flagella of *Chlamydomonas* (Figure 2.9a). A *Paramecium* is in effect a submarine driven by 5000 oars.

Ciliate protozoans range in length from about $25\,\mu m$ to 1 mm. Whatever

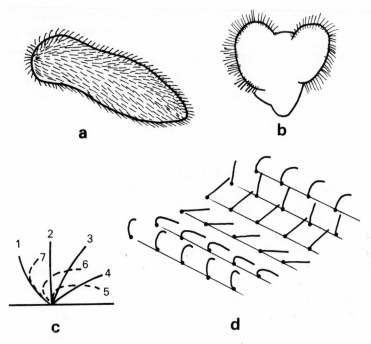

Figure 2.11   (a) *Paramecium*, a ciliate protozoan 0.25 mm long.
(b) Larva of a gastropod mollusc, 0.5 mm long.
(c) Diagram of a cilium beating.
(d) Diagram of a group of cilia on *Paramecium*, showing the metrachronal pattern of beating. (The cilia have been drawn unrealistically far apart for clarity.) This diagram is based on electronic flash photographs.

their size, nearly all of them swim at about 1 mm/s, ten times as fast as most flagellate protozoans (Sleigh and Blake, 1977). They have of course much larger "engines" than the flagellates. Compare the flagellate *Euglena viridis*, which has a cell body 50 μm long, with the ciliate *Tetrahymena*, which is about the same length. *Euglena* has a single flagellum 50 μm long. *Tetrahymena* has 500 cilia each 7 μm long, with a total length of 3.5 mm. It does not seem surprising that *Tetrahymena* is much faster than *Euglena*. It was argued for flagellates (p. 19) that speed could be increased very little by evolving more or longer flagella, because most of the power used for swimming is needed to propel the flagella themselves. This argument does not apply to ciliate protozoans because their manner of swimming is quite different.

    Some multicellular animals also swim by means of cilia. They include small flatworms and the larvae of many marine invertebrates (Figure 2.11b). Nearly all of them are less than 2 mm long but most ctenophores (sea gooseberries, etc.) are much larger.

Figure 2.11c shows a cilium of length l beating. In the power stroke it is extended but in the recovery stroke it is bent with its tip much closer to the cell body. The water immediately in contact with the cell body of a swimming ciliate moves forwards at the same speed as the cell body, u. The water at a distance l from the body moves backwards. It has been calculated that its velocity should be about $-u$, and it has been shown that this is the case by filming *Paramecium* swimming in a suspension of tiny polystyrene spheres (Blake and Sleigh, 1974).

This means that there is a steep gradient of velocity in the layer of water between the tips of the cilia and the cell surface. The cilia have to exert forces and do work against the viscosity of the water in this layer. If the cilia were longer, the gradient of velocity would be less steep and the forces and work correspondingly less. The forces would however act further from the cell surface, so the bending moments at the bases of the cilia would probably be about the same as before, for any particular swimming speed. Longer cilia would probably increase swimming speed very little, if at all. The speeds of ciliated animals may be limited by the number of cilia that can be conveniently fitted on to a unit area of cell surface, and the bending moment each cilium can exert.

The only ciliated animals capable of swimming much faster than 1 mm/s are ctenophores such as the sea gooseberry *Pleurobrachia*. This gelatinous animal has a diameter of 15 mm and swims at about 7.5 mm/s. Instead of separate cilia it has comb plates, bundles of about 10^5 very long cilia which are packed tightly together and beat in synchrony. This arrangement makes it possible to exert on the water the forces needed for relatively fast swimming, without requiring an excessive bending moment in any individual cilium.

On an ordinary ciliated animal, neighbouring cilia would collide if they did not beat at the same frequency and in an appropriate, orderly pattern (Machemer, 1972). Different animals use different patterns. The one used by *Paramecium* is shown in Figure 2.11d. Notice that adjacent lines of cilia are slightly out of phase with each other. The pattern of beating gives the impression of waves travelling over the animal, at right angles to the plane of beating. Coordinated movements like this, in which similar structures move at the same frequency but with regular phase differences, are described as showing metachronal rhythm. Another example is provided by the rays of an undulating fin (Figure 2.8a).

The metachronism of cilia seems to be maintained by hydrodynamic interaction between the cilia, rather than by any organizing centre in the cell. Indirect evidence for this comes from experiments with *Paramecium* treated with detergent to disrupt their cell membranes. These swim in solutions containing ATP, usually with a regular and apparently normal metachronal rhythm. Since the experiment prevented any nerve-like coordination involving the cell membrane it seems likely that the coordination is a purely mechanical effect.

Cilia normally beat in one direction, but the direction can be reversed. If *Paramecium* collides with an obstacle it reverses its cilia, backs off and advances again, usually in a different direction. It has been shown by experiments with detergent-treated *Paramecium* that the cilia beat as required for forward swimming when the concentration of calcium ions in the cell is very low, but reverse when the concentration is higher. It seems that the concentration is normally kept very low in the intact animal but that a collision makes the cell membrane temporarily permeable to calcium ions, which diffuse in. The event is very like the action potentials of metazoan nerves and involves similar electrical events, but calcium in *Paramecium* serves the function of sodium in a nerve.

Ducks

Ducks float on the water surface, rowing with their feet. Their movement produces waves like those produced by boats (Figure 2.10b), and this sets a practical limit to their speed of swimming (Prange and Schmidt-Nielsen, 1970). Water is given potential energy when it is raised in a wave. This energy must be supplied by the muscles of a duck, and by the engine of a ship. The power required for this becomes very large indeed at a speed of about $(0.16gl)^{1/2}$, where g is the acceleration of free fall and l is the length of the hull. Notice that at this speed, the Froude number, u^2/gl (p. 147) is 0.16. The Froude number arises in this context because wave formation involves interaction between gravity and inertia. Similar bodies, moving under the influence of gravity, move in dynamically similar ways only if their Froude numbers are equal.

A motorboat 5 m long reaches a Froude number of 0.16 at a speed of 3 m/s (10 km/h). A mallard duck (*Anas platyrhynchos*) has a "hull" about 0.33 m long and reaches a Froude number of 0.16 at a speed of 0.7 m/s. Mallard ducks on a pond swam at a mean speed of 0.5 m/s. Ducks tested in a laboratory experiment could swim faster, but they could not exceed 0.7 m/s for more than a few seconds.

Swimming with hydrofoils

Figure 2.12a shows a penguin swimming under water by beating its wings. The wing movements are quite unlike the leg movements of the water beetle in Figure 2.9b. The penguin is plainly not rowing.

Figure 2.12b shows what it is probably doing. The vertical sections through the wing are drawn at angles which seem from the film to be realistic. In the downstroke the wing has a positive angle of attack, so lift acts upward, but in the upstroke it has a negative angle of attack and lift acts downward (see Figure A.5b). The resultant of lift and drag acts forward and upward in the downstroke, forward and downward in the upstroke. The upward and downward components cancel each other out, in a complete cycle, so the net effect is forward thrust. Notice that the forward components of the forces derive solely from the lift, and that the

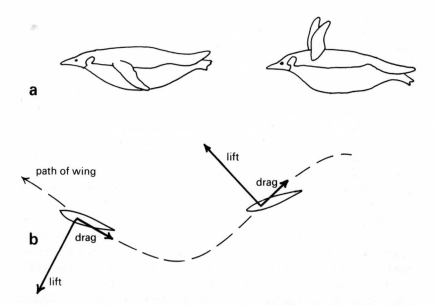

Figure 2.12 (a) Outlines traced from two frames of a film of an Emperor penguin (*Aptenodytes forsteri*) swimming. (From B. D. Clark and W. Bemis (1979) *J. Zool., Lond.* **188**, 411–428.)

(b) A diagram showing vertical sections through the penguin's wing, and the forces which probably act on it, at two stages in the cycle of swimming movements. The path of the section through the water is also shown.

drag tends to reduce them. This is a difference from rowing, in which the thrust is provided by drag on the oars.

Marine turtles swim like penguins, using their flippers as hydrofoils. Whales swim in essentially the same way but use the tail fluke instead of flippers. Similarly, some fish such as tunnies use their tail fins as hydrofoils. Whales have horizontal flukes which they move up and down but fish have vertical tail fins which they move from side to side.

There is no sharp distinction between undulatory and hydrofoil swimming in fish. They are simplified descriptions of the extremes in a range of actual swimming techniques. The undulatory description is plainly best for the eel, in which the whole body is thrown into obvious waves and the tail fin is merely a narrow strip round the posterior end of the body. The hydrofoil description, however, seems best for tunnies and whales because they bend little except in the posterior part of the body, which is narrow compared to the span of the tail fin (Hoar and Randall, 1978).

The span of a hydrofoil is its length from tip to tip (Figure A.5a), the chord is the distance from front edge to rear edge, and the aspect ratio is (span)/(mean chord). Equation A.19 shows that a hydrofoil of high aspect ratio can supply the same lift for less drag than one of the same area, but lower aspect ratio. The hydrofoils of penguins, turtles, whales and tunnies all have quite high aspect ratios. Also, they are all streamlined in section, like aeroplane wings. At the fairly high Reynolds numbers at which they work, streamlined sections give the best ratios of lift to drag.

A typical teleost has a fan-like tail fin, with separate rays joined by a flexible membrane. This makes possible undulating movements in which waves travel up or down the fin, which are useful in some manoeuvres. Such tails have rather low aspect ratios and are not streamlined in section. In evolving their hydrofoil tails, tunnies have lost the ability to undulate the tail fin.

Most animals which swim by means of hydrofoils are large and fast. Possible exceptions are swimming crabs (Portunidae) which may use their flipper-like hind legs as hydrofoils. King penguins (*Aptenodytes patagonica*) have been filmed swimming in a large aquarium at 3.4 m/s. Turtles (*Chelonia mydas*) swim at speeds up to 2 m/s. Dolphins and tunnies are considerably faster. The highest reliable speed for a dolphin was obtained by filming a trained *Stenella attenuata*, chasing a lure towed by a variable-speed winch across a lagoon in Hawaii. This animal, 1.9 m long, briefly reached 11 m/s. A tunny (*Euthynnus pelamis*) 0.5 m long has been filmed in a large tank reaching 9.5 m/s in a burst of speed. Speeds up to 21 m/s have been claimed for two related species but seem improbably high—these were obtained by recording the rate at which hooked fish pulled out a line.

It has been calculated that 2.6 kW would be needed to drive a rigid, well-streamlined body of the same size as *Stenella* through water at the maximum observed speed of 11 m/s. The power needed to propel *Stenella* is probably more than this because the action of the fluke must transmit some kinetic energy to the water. Also, the drag on a swimming dolphin is likely to be more than for an equivalent rigid body, because tail movements may thin the boundary layer as already described for undulating fish. Even the minimum estimate of 2.6 kW is remarkably high, for an animal of mass 53 kg. The maximum power output which a 70 kg human athlete can maintain for a few seconds, pedalling a bicycle ergometer and working a hand crank, is only about 1.4 kW.

It is quite possible that dolphin muscle can produce much more power than human muscle, but two attempts have been made to avoid this conclusion by suggesting that dolphins may suffer less drag than conventional streamlined bodies. Both are concerned with the pattern of flow in the boundary layer. A 1.9 m *Stenella* swimming at 11 m/s has a Reynolds number of 2×10^7. At this Reynolds number, most of the boundary layer of an ordinary streamlined body is turbulent (Figure A.4c), but the drag would be much less if it could be kept laminar. Aerofoils have been designed which maintain laminar flow over most of their surface at similar Reynolds numbers, and in these the thickest part is further back than in ordinary aerofoils. Dolphins have the thickest part of the body fairly well back, which might help to keep the boundary layer largely laminar. Also, it has been suggested that the skin of dolphins may have properties which tend to damp out turbulence. The suggestion was made following experiments with models covered by a spongy coating which seemed to reduce drag remarkably, but attempts to repeat the experiments did not confirm this result. In any case, there is evidence that the boundary layer of a swimming dolphin is mainly turbulent. The deceleration of *Stenella* has been measured in films showing it coasting along between bursts of swimming. The drag coefficient calculated from these decelerations is almost exactly the same as expected for a rigid streamlined body with a turbulent boundary layer.

Dolphins often leap out of the water in a manner which looks playful and wasteful of energy. It may actually save energy, in fast swimming, because very little drag acts on them while they are in the air. The energy required for a leap is less than the energy required to swim the same distance under water, if the dolphin is travelling fast enough.

Dolphins often swim near the bows of ships, apparently getting a cheap ride by placing themselves in the bow wave, in water which is being pushed along by the ship.

Jet propulsion

Cuttlefish and squids have their gills enclosed in a mantle cavity which has muscular walls (Figure 2.13a). When they enlarge the cavity, water is drawn in through a wide slit. When they contract it, water is squirted out through a tube called the funnel. A valve prevents inflow through the funnel, and the overlapping edges of the slit have a valve-like action preventing outflow through it. Gentle rhythmic contractions of the mantle muscles maintain one-way flow of water over the gills. This is sufficient for respiration, but much more vigorous contractions of the same muscles are used for jet propulsion (Trueman, 1980). A squid (*Loligo*) 0.2 m long has been filmed accelerating from rest to 2.1 m/s by a single contraction of its mantle. In this case its funnel was in the position shown in Figure 2.13 so the water was squirted out forwards, sending the animal backwards. Squids can also bend their funnels round to point backwards, and use them for forward jet propulsion (Figure 3.1c). Continuous swimming is done by repeated contractions. Some oceanic squid swim fast enough to leap from the water and land on the decks of ships.

A given amount of work will accelerate a squid to a higher speed, if it is used to eject a large mass of water at low speed, than if it is used to eject a

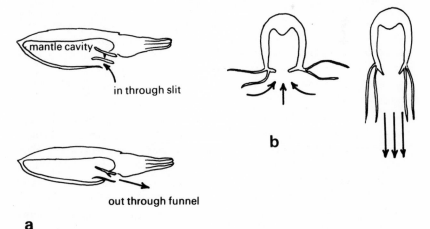

a

Figure 2.13 Diagrams of jet propulsion by (a) a squid and (b) a jellyfish.

smaller mass at high speed. The principle is the same as already proved for fish (equation 2.6). The tails of fish push on very large masses of water. Analysis of films of experiments like the ones shown in Figure 2.5 showed that each tail beat of an accelerating fish pushed on a mass of water averaging 3.5 times the mass of the fish. The masses are large because the tail effectively pushes on all the water which passes through the circle drawn round it in Figure 2.6. In contrast, a squid can only push, at each contraction, as much water as its mantle cavity will hold—it has been estimated for *Loligo* that this is 0.6 times body mass.

A few other animals also use jet propulsion but they are much less fast than squid. For example, dragonfly (*Aeschna*) larvae can swim at 0.1 m/s by squirting water from the rectum. Scallops (*Pecten* and *Chlamys*) swim rather erratically at speeds up to about 0.3 m/s, by rapidly opening and closing their hinged shells. As their shells close, water is squirted out in two jets on either side of the hinge. (Flaps of soft tissue prevent escape in other directions.) Consequently they swim with the hinge to the rear. Both dragonfly larvae and scallops seem to swim largely as a means of escape from predators. The swimming of jellyfish can also be regarded as jet propulsion, since water is squirted out from under the contracting bell (Figure 2.13b—Gladfelter, 1972).

CHAPTER THREE

BUOYANCY

Fresh water has a density of $1000\,kg/m^3$ but sea water, which contains much higher concentrations of dissolved salts, has a density of about $1026\,kg/m^3$. Many of the tissues of animals are considerably denser than either, because they contain proteins and (in some cases) inorganic crystals. For instance, the muscles and bones of fish have densities of about 1060 and $2000\,kg/m^3$, respectively. The soft parts of *Nautilus*, a swimming mollusc, have a mean density of $1060\,kg/m^3$ and the shell has a density of $2700\,kg/m^3$. Swimming animals without special adaptations for buoyancy are denser than the water they live in. For instance, dogfish (*Scyliorhinus*) have densities around $1075\,kg/m^3$, skipjack tuna (*Euthynnus*) $1090\,kg/m^3$ and squid (*Loligo*) $1070\,kg/m^3$. Among smaller swimmers, a copepod (*Labidocera*) has a density of $1080\,kg/m^3$. This chapter is about how dense animals avoid sinking, and about adaptations which reduce the densities of other animals so that they have no tendency to sink. A final section discusses the relative merits of different ways of avoiding sinking.

Dense animals

The simplest way for a dense animal to keep off the bottom is to swim upwards. Many small planktonic animals (for instance, copepods) have no other mechanism. Many larger animals swim horizontally, using fins as hydrofoils just as aeroplanes use wings to support them. Three examples are shown in Figure 3.1.

The weight of a body acts at its centre of mass. If the body is submerged in a fluid, an upthrust also acts on it, equal to the weight of an equal volume of fluid (by Archimedes' Principle). The upthrust acts at the geometrical centre of the body, which is generally close to the centre of

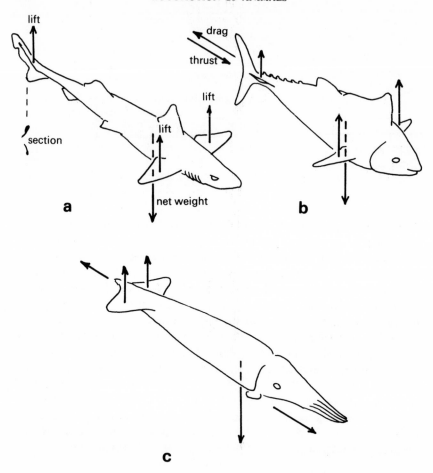

Figure 3.1 Sketches of (a) a shark, (b) a tunny and (c) a squid, all of which are denser than the water they swim in. Arrows represent forces that act on the body during swimming.

mass, but not precisely at it if one end of the body is denser than the other. It has been assumed for simplicity in Figure 3.1 that the upthrust acts at the centre of mass so that net weight (= weight − upthrust) can be shown as a single force acting at the centre of mass.

Sharks and tunnies (Figure 3.1a,b) swim with their pectoral fins extended with a positive angle of attack, so that upward lift acts on them (see Figure A.5b). These fins are anterior to the centre of mass, so if they were the only hydrofoils producing upward lift the fish would tend to tilt snout upwards. To balance the moments about the centre of mass, another source of lift is needed posterior to it. In sharks this is the tail, which is of the kind called heterocercal. The upper lobe of the tail is stiffened along its upper edge and the lower lobe along its lower edge, so the tail bends as it moves from side to side as shown in section under Figure 3.1a (Simons, 1970). The upper lobe tends to deflect water downwards, producing an upward force, but the lower lobe tends to deflect water upwards, and produce a downward force. The upper lobe produces the larger vertical component of force so the net effect is that the tail produces an upward force as well as forward thrust. It has been suggested that in tunnies, the posterior lifting surface may be the caudal peduncle (the slender segment of the body immediately anterior to the caudal fin).

Equilibrium of a swimming shark or tunny requires some particular upward force on each of the lifting surfaces. For instance, it has been estimated that a 1-kg dogfish (*Scyliorhinus*) needs upward forces of 0.13 N on the tail and 0.16 N on each pectoral fin (Alexander, 1965). If the angles of attack were constant, these forces would be about proportional to (speed)^2 (equation A.17). The angles of attack must therefore be adjusted to suit the speed of swimming. The various structures involved have muscles which seem capable of doing this. The same muscles can be used for vertical steering, when the fish changes its level in the sea.

If a hydrofoil is given too large an angle of attack it will stall. Consequently, there is a minimum speed below which a particular fin cannot provide the lift required of it. A shark or tunny that is denser than the water it lives in, cannot swim slower than this speed, which can be estimated from the areas of its hydrofoils. It is probably about 0.24 m/s for a dogfish and 0.6 m/s for a skipjack tuna (*Euthynnus*—Magnusson, 1978), both of mass 1 kg. These are close to the lowest speeds at which the fish swim in aquaria.

The squid *Loligo* (Figure 3.1c) has fins only at its posterior end. Upward lift acting on them would tend to tilt the squid nose down, but the thrust exerted by the jet from the funnel acts ventral to the centre of mass, tending to balance the moments.

The mandarin fish (*Synchiropus*) is a small, brilliantly coloured fish found on coral reefs in the south Pacific. Its density is 1150 kg/m^3 and it swims by undulating its pectoral fins. The undulations produce a forward, upward force, which serves both to propel the fish and to support its weight. A specimen 6 cm long used this method of swimming at speeds up to 0.2 m/s.

Low-density materials

Suppose an animal of volume V has density ρ but lives in water of lower density ρ_w. It can reduce its density to match the water by adding to its body a sufficiently large quantity of a substance even less dense than the water. Let this substance have density ρ_s and let the volume of it required be V_s. How big is V_s?

The original mass of the animal was $V\rho$ and the added mass of low-density substance is $V_s\rho_s$, giving a total mass $(V\rho + V_s\rho_s)$. The total volume is $(V + V_s)$. If the animal is to have the same density as the water

$$(V\rho + V_s\rho_s)/(V + V_s) = \rho_w$$

which can be rearranged to give

$$V_s/V = (\rho - \rho_w)/(\rho_w - \rho_s) \qquad\qquad 3.1$$

To see what this implies, consider an animal with $\rho = 1075\,\text{kg/m}^3$ (a typical density for fishes and squids that have no adaptations for buoyancy). Table 3.1 shows values of V_s/V for such an animal, for various low density substances and for both fresh and sea water.

Fat is less dense than water so it could in principle be used to match the densities of animals to the water they live in. The table shows that enormous quantities would be required. A marine animal would need enough fat to increase its volume by about 51 %, and a freshwater animal would need far more fat even than this. Few animals, if any, have so much fat. Other organic compounds with lower densities are more effective.

Though dogfishes and many other sharks are considerably denser than sea water, a few sharks have almost exactly the same density as sea water (Corner, Denton and Forster, 1969). They include sharks such as

Table 3.1 The volumes of various substances which must be added to the body of an animal of initial density $1075\,\text{kg/m}^3$, to match its density to fresh water or sea water. These volumes have been calculated from equation 3.1 and are expressed as fractions of the initial volume (i.e. as V_s/V).

	fresh water $\rho_w = 1000\,\text{kg/m}^3$	sea water $\rho_w = 1026\,\text{kg/m}^3$
fat, $\rho_s = 930\,\text{kg/m}^3$	1.07	0.51
squalene and wax esters $\Big\} \rho_s = 860\,\text{kg/m}^3$	0.54	0.30
gas, $\rho_s \simeq 0$	0.08	0.05

Centroscymnus which live near the sea bottom at depths of the order of 1 km, and also the huge basking shark (*Cetorhinus*) which lives at the surface. These sharks have enormous livers containing very large quantities of the hydrocarbon squalene, which is less dense than fat. For instance, the liver of *Centroscymnus* occupies 30% of the volume of the body. Eighty per cent of this liver is oil, largely (probably mainly) squalene.

Many bony fishes have gas-filled swimbladders, but a few which do not nevertheless have densities very close to sea water. They include the coelacanth *Latimeria* and some of the small oceanic fish called lantern fishes (Myctophidae). These fish contain large quantities of oily substances, largely wax esters of about the same density as squalene. Their flesh is very oily, and there are also oily deposits in the body cavity. Lantern fishes that accumulate wax esters as adults, have swimbladders as juveniles. The coelacanth has a peculiar fatty organ which seems to be a degenerate swimbladder. Wax esters seem to have replaced the swimbladder, in the course of the evolution of these fish.

Some animals rely on low-density body fluids for buoyancy. Solutions of different salts with the same osmotic concentration have different densities. For instance, solutions of sodium chloride, sodium sulphate and ammonium chloride, of the same osmotic concentration as sea water, have densities of about $1018 \, kg/m^3$, $1040 \, kg/m^3$ and $1007 \, kg/m^3$, respectively. The densities of body fluids can be reduced by eliminating dense ions such as sulphate or incorporating light ones such as ammonium.

Sodium and chloride are much the most plentiful ions in sea water. The fluids in the body cavities of some deep-sea squids have about the same osmotic concentration as sea water but contain far more ammonium than sodium (Denton, 1974). In the case of *Helicocranchia*, this body fluid has a density of only $1010 \, kg/m^3$. The density of the rest of the body is about $1050 \, kg/m^3$ and the intact animal has about the same density as sea water. By equation 3.1, the volume of low-density fluid must be 1.5 times the volume of the rest of the body. The body cavity is enormously swollen and the animal looks curiously bloated.

The ammoniacal squids are seldom seen, for they swim mainly at depths of several hundred metres. They must nevertheless be common. Examination of the stomach contents of sperm whales (*Physeter*) shows that ammoniacal squid are their principal food.

Jellyfish (Scyphozoa) also gain buoyancy by manipulating their ion content (Bidigare and Bigg, 1980). They consist largely of mesogloea, which is a very dilute protein gel. If this gel were made up in sea water it would inevitably be more dense than sea water itself. In fact it is slightly less dense; this seems to be because it contains less sulphate than does sea water, and correspondingly more of other, lighter ions. (Only a small reduction of density can be obtained in this way because sea water itself contains only a little sulphate.) The individual cells of marine jellyfish are denser than sea water and some jellyfish are, overall, denser than sea water, but others, such as *Pelagia*, contain very large volumes of mesogloea and have about the same density as the sea water they live in. As well as jellyfish, sea gooseberries (Ctenophora) and some siphonophores gain buoyancy from mesogloea with a low sulphate content. Similarly, *Pyrocystis noctiluca*, a huge non-swimming flagellate protozoan, contains a vacuole filled with a solution containing hardly any sulphate. The vacuole occupies 95% of the volume of the cell and gives the animal a density slightly less than sea water.

Gas-filled floats

Gas-filled floats are particularly effective as buoyancy organs, because gases have such low densities. Table 3.1 shows that much smaller volumes of gas than of fats, squalene or wax esters will match the densities of animals to the water they live in.

Most teleost fish have a swimbladder, a gas-filled sac in the body cavity (Figure 3.2b,c—Alexander, 1966; Blaxter and Tytler, 1978). It is usually filled mainly with nitrogen and oxygen, with sometimes a good deal of carbon dioxide. It usually has the right volume to match the density of the fish almost exactly to the water. Fish floating in aquaria, hardly moving a fin, show how precise that match can be.

There are various difficulties associated with swimbladders. If a fish with a swimbladder swims deeper in the water, its swimbladder is compressed by the increased pressure, so the density of the fish increases. If the fish moves nearer to the surface, its density decreases. There is only one depth at which its density matches the water. Even at this depth its equilibrium is unstable, like the equilibrium of a pencil balanced on its point. Any disturbance which makes the fish rise slightly in the water moves it to a lower pressure, so its swimbladder expands and it tends to rise further. Any disturbance which makes it sink slightly causes an increase of density leading to further sinking. Because of this instability, a fish floating in an aquarium is seldom absolutely motionless, but has to make occasional fin movements to preserve its equilibrium.

Swimbladders probably evolved from the lungs of early bony fish, but a lung is a poor swimbladder except near the surface. Imagine a fish with

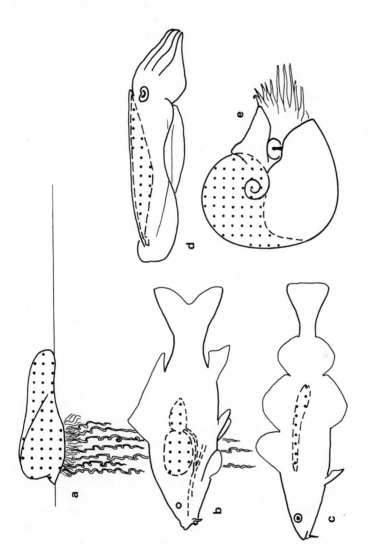

Figure 3.2 Sketches of animals with gas-filled floats. Internal structures are indicated by broken lines and the position of the gas is shown by stippling. (a) The Portuguese man-o'-war (*Physalia*); (b) a goldfish (*Carassius*); (c) a cod (*Gadus*); (d) a cuttlefish (*Sepia*) and (e) *Nautilus*.

both gills and a lung. The blood comes into close contact with the water at the gills and with air in the lung, so gases can diffuse to or from the blood at both places. (The respiratory functions of gills and lungs depend on this.) The pressure of the gas in the swimbladder of a submerged fish is greater than one atmosphere, because of hydrostatic pressure. The total of the partial pressures of dissolved gases in the sea is 1 atm or a little less at all depths, because the sea tends to equilibrate with the atmosphere. (It is apt to be less than 1 atm because oxygen is used for respiration by marine organisms.) Consequently there is a tendency for gases to diffuse from the lung to the blood, and (at the gills) from the blood to the water. The lung will gradually deflate. If the fish were near the surface it could make the loss good by gulping air, but it cannot do this while swimming at depth.

Diffusion losses have been reduced by the evolution of swimbladders, which are very unlike lungs. Lungs have complex arrangements of alveoli which give them very large internal surface areas for diffusion, but most swimbladders are smooth internally. Lungs are well supplied with blood which is brought very close to the air but swimbladders have very sparse blood supplies, except to specialized regions concerned with addition and removal of gas. Many swimbladders have a layer in their walls which is made remarkably impermeable to gases by closely packed crystals of guanine.

Teleosts have gas glands in their swimbladders, capable of secreting gases if necessary against large differences of partial pressure. (The mechanism is described in books on comparative physiology.) Thus gas lost by diffusion is replaced, and extra gas is added as required if the fish moves to a greater depth. Teleosts get rid of excess gas when adjusting to reduced depths, either by releasing gas from their mouths or by absorbing it into the blood. The swimbladders of primitive teleosts have a connection to the mouth (like a trachea, Figure 3.2b) but the connection has been lost by more advanced teleosts (Figure 3.2c).

The processes of secreting gas into the swimbladder and reabsorbing it into the blood are slow. In many experiments, the swimbladder has been deflated and the time taken for refilling observed. The time is always several hours: the shortest known to me is 4 h, by the bluefish *Pomatomus*. The pressure in the sea increases by about 1 atm for every 10 m descent. The volume of gas (measured at atmospheric pressure) required to inflate a swimbladder to volume V is therefore V at the surface, $2V$ at 10 m, $3V$ at 20 m and so on. A fish requiring 4 h to secrete a volume V cannot adjust to increases of depth at rates faster than 2.5 m/h.

Some of the disadvantages of swimbladders are avoided by the gas-filled shells of some cephalopod molluscs, including cuttlefish and *Nautilus* (Fig. 3.2d,e—Denton, 1974). The gas is enclosed in a rigid shell, not a flexible sac. It is below atmospheric pressure, no matter how deep the animal is living. The difference can be illustrated by comparing a fish such as a hake, *Merluccius* (which has no duct connecting the swimbladder to the mouth) with *Nautilus*, both caught at a depth of 100 m where the pressure is 11 atm. The hake arrives at the surface with its swimbladder enormously swollen or burst, possibly forcing the viscera out through the mouth. The *Nautilus* arrives at the surface unharmed, with about the same density as the water. If it is held under water and a hole is bored in its shell, gas does not bubble out: rather, water is sucked in. A typical *Nautilus* had gas pressures ranging from 0.4 atm (in the newest chamber of its shell) to 0.8 atm (in older chambers).

Cephalopods do not secrete gas into their shells. Instead they withdraw water (probably by an osmotic mechanism) leaving a partial vacuum into which gases tend to diffuse from the body fluids. Consequently the partial pressures of gases in the shell do not exceed the partial pressures of the same gases dissolved in the sea water.

The rigid shell does not change volume appreciably with depth, but it may be crushed if the animal swims too deep. Measurements have been made of the pressures the shells can withstand by exposing them to high pressures in pressure chambers. Cuttlefish shells collapsed at 24 atm, but cuttlefish do not seem to live deeper than 150 m, where the pressure is 16 atm. *Nautilus* shells shattered at about 65 atm, but *Nautilus* does not seem to live deeper than 500 m where the pressure is 51 atm.

Since the shell has to be strong, quite a lot of dense material is needed to build its wall. The shell wall of *Nautilus* is made of similar material to the shells of other molluscs, with a density of 2700 kg/m³. Even with its chambers completely full of gas, the density of the shell is no less than about 910 kg/m³, which is close to the density of typical fats (Table 3.1). The shell of the cuttlefish is the "cuttlebone" which is often given to cage birds to peck. It is a stack of gas chambers with thin walls held apart by pillars. It is much more lightly built than the shell of *Nautilus*, with an overall density of about 600 kg/m³. The density of the cuttlefish without its cuttlebone is about 1067 kg/m³, so by equation 3.1 the required volume of cuttlebone is about 10 % of the volume of the animal. This is double the usual volume for a teleost swimbladder but much smaller than the shell of *Nautilus*.

Gas-filled floats are also possessed by siphonophores, which are colonial hydroids. Each colony has a float containing oxygen, nitrogen and (surprisingly) carbon monoxide. The Portuguese man-o'-war has a very large float, and floats with most of it above the surface (Figure 3.2a). Other siphonophores live deeper in the oceans and have much smaller floats or none at all: the ones without floats owe their buoyancy to the exclusion of heavy ions from their mesogloea.

The lungs of aquatic tetrapods and the plastrons of some aquatic insects are gas-filled respiratory organs which inevitably reduce the animal's density. The air in the lungs enables people to float. Some of the terrapins (freshwater Chelonia) adjust the volume of air in their lungs to match their density to that of the water they swim in.

Buoyancy and ways of life

Different buoyancy mechanisms are associated with different ways of life, in fish and cephalopods. Fish that spend a lot of time lying on the sea bottom generally have no buoyancy organ and are denser than water. Examples are plaice (*Pleuronectes*, a teleost without a swimbladder) and rays (*Raia*, selachians). *Octopus* also lives mainly on the bottom, has no buoyancy organ and is denser than sea water. High density can be a positive advantage for an animal resting on the bottom: only if it is denser than the water will there be frictional forces to stop it sliding around. For fish which live on the bottom in surf or fast streams, high density may not provide sufficient anchorage. Some such as the lumpsucker (*Cyclopterus*, a shore fish) have evolved suckers which they use to attach themselves to rocks.

As well as fish which rest on the bottom and swim seldom, fish which swim perpetually at high speeds tend to be denser than the water. Mackerels and tunas (Scombridae) never stop swimming, day or night, when kept in aquaria. Many of them have no swimbladder and have densities between 1070 and 1090 kg/m^3. Many swim quite fast (and have to do so, to get the upward lift they need from their fins). For instance, bullet mackerel (*Auxis rochei*), 0.3 m long, swim at an average speed of 0.7 m/s. There are dense sharks including the blue shark (*Prionace*) that seem to swim perpetually. Squids such as *Loligo* also swim perpetually (at least in aquaria) and are denser than sea water.

Gas-filled floats have not been evolved by sharks, but most teleost fish have swimbladders. The principal exceptions are the bottom-livers, the fast swimmers and some deep-sea teleosts. Cuttlefish and *Nautilus* also have gas-filled floats. Cuttlefish kept in aquaria swim around all night and bury themselves (if sand is available) by day. They spend a lot of time swimming slowly around by undulating their fins, like some teleosts with swimbladders.

There seems to be a tendency for animals living at substantial depths in the oceans to have non-gaseous buoyancy aids. *Centroscymnus* and similar sharks, *Latimeria*, lantern fishes and ammoniacal squid all spend at least part of their time at depths of several hundred metres, but in other respects their ways of life are very different. *Centroscymnus* live near the bottom at depths of the order of 1 km. *Latimeria* are caught near the bottom at depths of 400 m or less. Ammoniacal squid live a few hundred metres from the surface in water which extends much deeper. Lantern fishes spend the day at similar depths but spend the night much nearer the surface, and some can be netted at the surface at night.

There is, however, no simple relationship between buoyancy mechanisms and depth. Some lantern fishes have swimbladders and some siphonophores which live at similar depths (joining them in their vertical migrations) have gas-filled floats. *Nautilus* lives as deep as many ammoniacal squid and the related *Spirula* (much smaller, but with a gas-filled shell) has been caught at 1200 m. Macrourids and other teleosts which live near the bottom, much deeper in the oceans, have swimbladders. Basking sharks and jellyfish live near the surface but have non-gaseous buoyancy aids.

Advantages of different buoyancy aids

This section seeks explanations for the different buoyancy aids of fishes with different ways of life. It is based on the hypothesis that evolution tends to favour whichever mechanism is most economical of energy. It will be convenient to base discussion on a fish of specified size, though the qualitative conclusions will apply to fish of all sizes. The chosen example is an imaginary marine fish with the same dimensions as the 0.28 m trout discussed earlier (p. 23). Its density without buoyancy adaptations is $1075 \, kg/m^3$, as assumed in Table 3.1. The energy cost of swimming will be discussed, for each of several means of avoiding sinking. Estimates will be made of the power needed to prevent sinking, in excess of the "basic" power which would be needed if its density could be reduced to match that of the water without any change of volume.

Consider swimbladders and wax esters (or squalene) first. The required volumes are 5% and 30%, respectively, of the volume of the fish without buoyancy organs (Table 3.1). Since surface area is proportional to (volume)$^{2/3}$, for geometrically similar fish, an adequate swimbladder adds

3% to the surface area of a fish ($1.05^{2/3} = 1.03$), and wax esters add 19%. The power required for swimming at given speed is presumably proportional to surface area. Hence power needed to prevent sinking is 3% of the basic power for a swimbladder, and 19% for wax esters. The basic power has been calculated from observations of the tail movements of 0.28 m trout. Thus the lines for swimbladder and wax esters, in Figure 3.3, have been calculated.

Now consider a fish which depends on hydrofoils to prevent sinking. If its mass and density are as given under Figure 3.3, its net weight in sea water is 0.1 N. The pectoral fins of a fish of this size would have a chord of about 2 cm, so if the fish swam at speeds of the order of 1 m/s the Reynolds

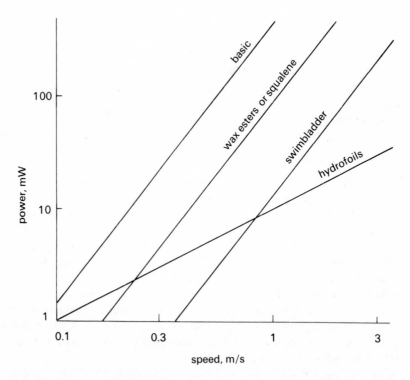

Figure 3.3 Estimates of mechanical power required for swimming by a marine fish which without buoyancy organs, has length 0.28 m, mass 0.22 kg and density 1075 kg/m³. The basic power required for swimming is shown, and also the additional power needed to prevent sinking by various means. The density of sea water is assumed to be 1026 kg/m³.

numbers of its fins would be of the order of 2×10^4. At such Reynolds numbers, the drag on a hydrofoil of optimum area is likely to be about one tenth of the lift it is designed to produce, so the extra drag on the fish due to the hydrofoils can be estimated as 0.01 N. The corresponding extra power is this drag multiplied by the speed, and is shown in Figure 3.3. (It has been assumed that the area is adjusted to the optimum for each speed and also that the tail works with 100% efficiency, which is accurate enough for this rough calculation.)

Figure 3.3 suggests that a swimbladder is always more economical than wax esters or squalene, as a means of preventing sinking. It is more economical than hydrofoils at low speeds but less economical at high speeds. This tallies with the observation that many teleosts with swim-bladders spend a lot of time stationary or swimming slowly while Scombridae without swimbladders swim perpetually, and rather fast. *Auxis* (density 1086 kg/m^3) is about the same size as the fish on which the figure is based and swims at about 0.7 m/s, which may be fast enough for hydrofoils to be more economical than a swimbladder.

Swimbladders have not been evolved by sharks, which depend either on hydrofoils or on squalene. Figure 3.3 suggests that squalene is the more economical option only at low speeds, but it is not clear from the figure (which is based on a small fish) what the critical speed would be for a large shark. The very large basking shark, which depends on squalene, swims at the remarkably low speed of 1 m/s. The speeds of deep-sea sharks which depend on squalene are not known.

Figure 3.3 takes no account of depth changes such as lantern fishes make every day (Alexander, 1972). These may destroy the advantage of a swimbladder over wax esters. It has been shown, by echo-sounding com-bined with direct observation from a research submarine, that many lantern fishes spend the night near the surface and the day at depths around 300 m. Among the lantern fishes caught at the surface at night, those with swimbladders have about enough gas in them to match their densities to the water. If these fish swim to 300 m by day, the extra pressure must compress the gas to 1/30 of its night-time volume so that it con-tributes hardly anything to their buoyancy. It seems inconceivable that they could secrete gas fast enough to compensate for the depth change, so they presumably have to use their fins as hydrofoils to prevent sinking during the day. The energy cost of avoiding sinking, for these fish, is not the cost shown by the swimbladder line in Figure 3.3 but the mean of the night-time cost (given by the swimbladder line) and the daytime cost (given by the hydrofoil line). This could be more than the cost using wax esters (which are effective at all depths), but only at very low speeds. Figure 3.3 is not directly applicable to the lantern fishes, which have masses of only a few grams, so it is not clear how low the critical speed would be.

Lantern fishes feed at night. Observations from the research submarine show that species with wax esters spend much of their time by day hanging motionless in the water. They could not do this if they depended on upward lift from hydrofoils. If the daytime choice is between hanging motionless (for species with wax esters) and swimming perpetually (for species with swimbladders), wax esters seem likely to offer the more economical option. (There is, however, an unexplained observation of a species possessing a swimbladder hanging motionless by day!)

There is a further energy cost associated with squalene and wax esters, which has so far been ignored. This is the cost of synthesizing more material, as the animal grows. The cost is apt to be large because these compounds are required in large quantities and have large heats of combustion. The costs of growth of a swimbladder wall or of large hydrofoil fins are negligible by comparison.

Consider a fish of basic mass m, plus an additional mass qm of a low density compound of heat of combustion H. Let the growth rate be k so that the fish is adding to its basic mass at a rate km. To keep the proportion of buoyancy compound in its body constant, it must invest energy in the buoyancy compound at a rate $kqmH$. This energy would otherwise have been available to the muscles which would have used it with efficiency η, doing work at a rate $\eta kqmH$. The energy cost of accumulating the buoyancy compound, expressed as an equivalent mechanical power, is therefore $\eta kqmH$.

For squalene and wax esters, H is probably about 40 MJ/kg. The fish is assumed to consist of a basic volume V of tissues of density $1075\,\text{kg/m}^3$ plus a volume $0.3V$ of squalene or wax esters of density $860\,\text{kg/m}^3$ (Table 3.1), whence q is 0.24. Its basic mass m is assumed to be $0.22\,\text{kg}$ (Figure 3.3). The experiment illustrated in Figure 2.7c indicated muscle efficiencies η up to about 0.16. Fish of mass $0.22\,\text{kg}$ are likely to have growth rates in the range 0.1 to 1 per year, 3×10^{-9} to $3 \times 10^{-8}/\text{s}$. Hence the power $\eta kqmH$ is likely to be between 1 and 10 mW. Compare this with the power requirements shown in Figure 3.3. A fish with squalene or wax esters which did not grow too fast and spent a lot of time hanging motionless in the water, might well be more economical of energy than one which depended on hydrofoils and had to keep swimming. It might also be more economical than a fish with a swimbladder that descended by day to depths where the swimbladder was useless. The cost of growth probably does not destroy the advantage of wax esters, for those lantern fishes that depend on them.

It might be supposed that these fish could live even more economically, by relying on swimbladders and staying near the surface day and night. This may not be the case, because surface water is generally warmer than deeper down, and the metabolic rates of fishes are higher at higher temperatures.

Squalene or wax esters, in adequate quantities for buoyancy, add sub-stantially to the body masses of fish. Any particular force will give the body a smaller acceleration than if it had a swimbladder instead. Squalene and wax esters must reduce the ability of fish to accelerate, brake and turn sharply.

However impermeable the wall of a swimbladder, it must lose some gas by diffusion. The energy cost of replacing this gas is difficult to estimate because the efficiency of gas secretion is not known. A rough calculation suggests it is likely to be small, compared to the costs estimated in Figure 3.3, except possibly for small fish at depths greater than 1 km. Teleosts with gas-filled swimbladders are common near the ocean floor, even at depths of several kilometres, but few if any of them are very small.

In conclusion, swimbladders are probably the most economical means for teleosts to prevent themselves from sinking, unless they swim fast or make large vertical migrations. Hydrofoils are better for teleosts that swim perpetually at high speeds, and wax esters may be best in some circum-stances for those that make large vertical migrations. Squalene is probably more economical than hydrofoils for sharks which swim slowly and do not grow too fast, but hydrofoils are better for other sharks.

CHAPTER FOUR

FLIGHT

Gliding

Since gliding is simpler than powered flight it will be discussed first (see also Pennycuick, 1975).

Figure 4.1a is a diagram of a gliding bird, whose wings are spread but held motionless relative to the trunk. Its velocity relative to the air is u, at an angle θ to the horizontal. It will be assumed to have no acceleration, which implies that it is in equilibrium under its weight mg and the aerodynamic forces on its body. The resultant aerodynamic force has a lift component L (at right angles to the direction of motion) and a drag component D (backwards along the direction of motion). For equilibrium

$$L = mg \cos \theta \qquad 4.1$$

$$D = mg \sin \theta \qquad 4.2$$

The standard equation for the lift on an aerofoil of plan area S_p in air of density ρ is

$$L = \tfrac{1}{2}\rho S_p u^2 C_L \qquad 4.3$$

where C_L is the lift coefficient (equation A.17). If the gliding angle θ is small, as it usually is for gliding animals, $\cos \theta \simeq 1$ and equations 4.1 and 4.3 give

$$mg = \tfrac{1}{2}\rho S_p u^2 C_L$$
$$u = (2N/\rho C_L)^{1/2} \qquad 4.4$$

where N ($= mg/S_p$) is the wing loading, the weight supported by unit area of wing. A bird can alter C_L by adjusting the angle of attack of its wings, so it can glide at various speeds. It cannot however increase C_L above a maximum value $C_{L\max}$ which probably lies between 1 and 1.5 for nearly

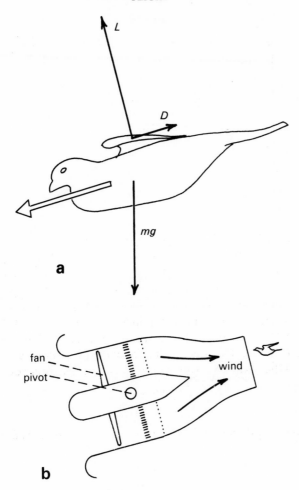

Figure 4.1   (a) Diagram of a gliding bird.
(b) Tilting wind-tunnel used for experiments with gliding birds.

all gliding animals (appendix, p. 151). There is therefore a minimum gliding speed u_1 given by

$$u_1 = (2N/\rho C_{L\,\mathrm{max}})^{1/2}$$ 4.5

In some circumstances it may seem best to glide as slowly as possible, for instance in coming in to land. In others it may be better to glide at as

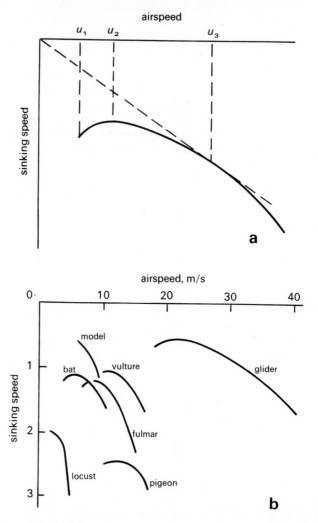

Figure 4.2 Graphs of sinking speed ($u \sin \theta$) against airspeed (u) for gliding animals and aircraft. Sinking speed is conventionally plotted downwards on graphs like this because sinking is downward movement.

(a) A schematic graph showing the minimum gliding speed u_1, the speed u_2 at which the sinking speed is least, and the speed u_3 at which the gliding angle is least.

(b) Graphs of minimum observed sinking speed against airspeed for a full-sized glider, a model glider, locusts (*Locusta*, about 2 g), a bat (*Rousettus*, 120 g), pigeons (*Columba*, 400 g), fulmars (*Fulmarus*, 700 g) and a vulture (*Coragyps*, 1.8 kg). Data from C. J. Pennycuick (1971) *J. exp. Biol.* **55**, 833–845 and P. S. Baker and R. J. Cooter (1979) *J. comp. Physiol.* **A131**, 89–94.

small an angle θ as possible, to travel as far as possible for given loss of height. In yet others it may be best to minimize the sinking speed $u \sin \theta$ (the rate of loss of height relative to the air) so as to remain airborne as long as possible. Equation 4.2 gives

$$\sin \theta = D/mg \qquad 4.6$$

$$\text{and} \quad u \sin \theta = uD/mg \qquad 4.7$$

The drag D has two components, the profile drag $(\frac{1}{2}\rho S_p u^2 C_{DO})$ which would act even if no lift were being produced and the induced drag (about $m^2 g^2 / 2\rho S_p u^2 A$) associated with the production of lift (equation A.19). A is the aspect ratio. It is assumed, realistically, that the lift is approximately mg. Hence

$$\sin \theta = (\rho C_{DO} u^2 / 2N) + (N/2\rho A u^2) \qquad 4.8$$

$$u \sin \theta = (\rho C_{DO} u^3 / 2N) + (N/2\rho A u) \qquad 4.9$$

The right-hand side of each of these equations consists of two terms. The first term is large when u is large and the second is large when u is small. There is therefore a particular speed u_3 at which $\sin \theta$ has a minimum value and another speed u_2 at which $u \sin \theta$ has a minimum value. These speeds are shown in Figure 4.2a. By differentiating equations 4.9 and 4.8, it can be shown that

$$u_2 = (N^2 / 3\rho^2 A C_{DO})^{1/4} \qquad 4.10$$

$$\text{and} \quad u_3 = (N^2 / \rho^2 A C_{DO})^{1/4} = 1.32 u_2 \qquad 4.11$$

Equations 4.5, 4.10 and 4.11 show that u_1, u_2 and u_3 are all proportional to $N^{1/2}$: high wing loadings are suitable for faster flight than low ones. For geometrically similar animals of different sizes, wing areas would be proportional to (body mass)$^{2/3}$ so wing loading (weight/area) would be proportional to (body mass)$^{1/3}$. Each of the speeds u_1, u_2 and u_3 would be proportional to (body mass)$^{1/6}$. Animals of different sizes are generally not geometrically similar, but the range of sizes of flying animals is so large that large differences of wing loading seem inevitable (Figure 4.3). Large gliding animals (such as albatrosses) have much larger wing loadings than small ones (such as dragonflies) and glide correspondingly faster.

The minimum gliding angle θ_{min} can be obtained from equations 4.8 and 4.11

$$\sin \theta_{min} = (C_{DO}/A)^{1/2} \qquad 4.12$$

Wing loading does not appear in this equation. Large animals can nevertheless glide at shallower angles than small ones because they have

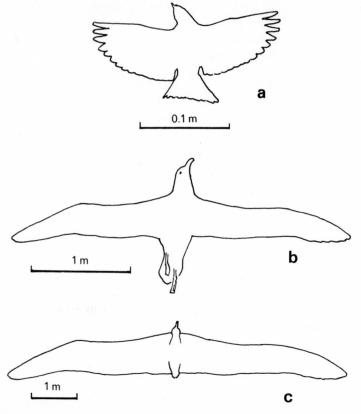

Figure 4.3 Outlines of (a) a sparrow, *Passer domesticus* of mass 30 g and wing area 100 cm², and (b) an albatross, *Diomedea exulans*, of mass 8.5 kg and wing area 0.62 m².
(c) An imaginary 8.5 kg albatross with wings of the size required to give it the same wing loading as the sparrow.

smaller values of C_{DO}. Most of the profile drag on aerofoils is friction drag, and friction drag coefficients fall as Reynolds number increases (equation A.16).

Equation 4.12 also shows that animals with large aspect ratios (A) should be able to glide at shallower angles than ones with small aspect ratios. An albatross (Figure 4.3b) should be able to glide at a shallower angle than a vulture (Figure 4.4).

Figure 4.2b shows the gliding performance of various animals. The data for locusts were obtained by analysing films of natural swarms, taken on occasions when there was little or no wind. The data for fulmars were derived from careful observations of birds gliding near a cliff, and of the local air movements. The other birds and the bat were trained to fly in a tilting wind tunnel, so as to be stationary relative to the laboratory (Figure 4.1b). If the tunnel was tilted steeply enough they could glide, but otherwise they had to flap their wings. The minimum angle at which they could glide was found for each of several wind speeds.

Figure 4.2b shows, as expected, that large gliding animals such as vultures glide best at higher speeds than small ones such as locusts (i.e. they have higher values of u_2 and u_3). Glider aircraft glide faster still. Minimum gliding angles for the animals range from 5° for the vulture to 30° for the locust. The pigeon glides much less well than the other birds and the bat. Flying phalangers (*Petaurus*, not included in the figure) glide even worse (Nachtigall, 1979). They are squirrel-like marsupials, and use a flap of skin between the fore and hind legs for gliding. Their minimum gliding angle is high (about 20°) because their aspect ratio is low and also because the body is so large relative to the "wing" area that the drag coefficient C_{DO} must be large—see equation 4.12.

Wing loading has been treated as a constant so far, but birds can alter it by extending or partly folding their wings (Figure 4.4a,b). Suppose a bird is attempting to glide at a particular speed u with as small a gliding angle as possible. It can be shown by differentiating equation 4.8 with respect to N that the optimum wing loading is

$$N_{opt} = \rho u^2 (C_{DO} A)^{1/2} \qquad 4.13$$

At speeds above u_3 (Figure 4.2a) this optimum is greater than the value for fully spread wings, so the bird does best by partly folding its wings as in Figure 4.4b. (This alters the aspect ratio and probably the drag coefficient, as well as increasing the wing loading, so this argument is unrealistically simple, but the conclusion is sound). Gliding birds reduce their wing areas considerably at high speeds but bats cannot alter theirs much without spoiling the aerodynamic properties of the wings. Folding a bat wing slackens the membrane between the digits. At very low speeds, birds often supplement their wing area by spreading the tail (Figure 4.4a).

If the bird shown in Fig. 4.1a is gliding at constant speed, it must be in equilibrium. Its weight must be in line with the resultant of lift and drag. Just as the weight acts like a single force at the centre of mass, the aerodynamic forces act like a single force at a point called the centre of pressure. For stable equilibrium, the centre of pressure must be vertically over the centre of mass. Birds can therefore tilt their bodies to suit different

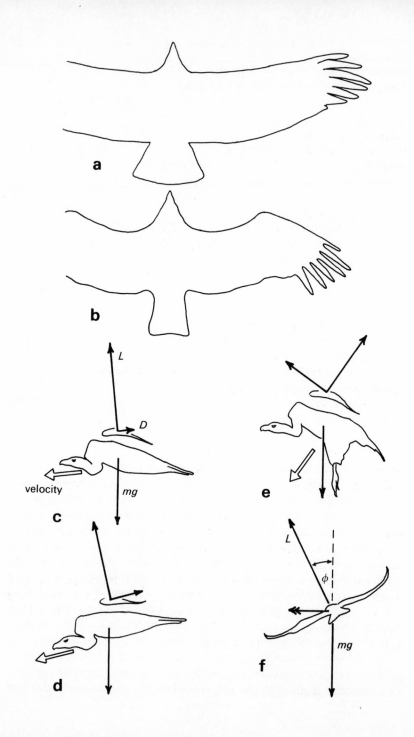

gliding angles by moving their wings (and thus the centre of pressure) forward (Figure 4.4c) or back (Figure 4.4d).

The lines in Figure 4.2 show the minimum possible sinking speed, for each forward speed. Birds can glide more steeply, at any forward speed, by using their feet as brakes (Figure 4.4e). For equilibrium, the resultant of lift and drag must be vertical, so an increase in drag makes the gliding angle steeper.

A bird that is circling at constant speed is not in equilibrium, but has an acceleration towards the centre of the circle (equation A.6). To turn, birds (like aeroplanes) bank, so that the resultant of weight and aerodynamic forces acts horizontally towards the centre of the circle (Figure 4.4f). Birds make themselves bank by rotating one or both wings about their long axes, so that one wing has a larger angle of attack than the other and obtains more lift.

Soaring

An animal gliding in still air cannot remain airborne indefinitely, but must either sink to the ground or resort to flapping flight. There are however natural air movements in the atmosphere that can be used to prolong gliding flight. Exploitation of these is called soaring.

Thermal soaring

One technique of soaring uses thermals, which are upward air movements due to irregular heating of the ground by the sun (Pennycuick, 1972). Dry, dark areas of ground and hillsides facing the sun are apt to get warmer than the surrounding ground. The air over them is heated and rises, either as a continuous column (a dust devil) or as a succession of vortex rings which rise at intervals. Thermals are used by glider pilots and also by birds, which can rise by circling in the thermal. The aircraft or bird sinks relative to the air (as it must, when gliding) but it rises relative to the ground if the air is rising fast enough. Thermals over the East African plains, used by vultures, contain air rising at (typically) about 4 m/s. Since vultures have minimum sinking speeds much less than this (Figure 4.2b) they can gain height by circling in such thermals. To travel across country they glide from thermal to thermal (Figure 4.5a). Glider pilots find

Figure 4.4 Sketches of vultures in gliding flight (based on illustrations by J. McGahan (1973) *J. exp. Biol.* **58**, 225–237, and C. J. Pennycuick (1971) *J. exp. Biol.* **55**, 39–46).

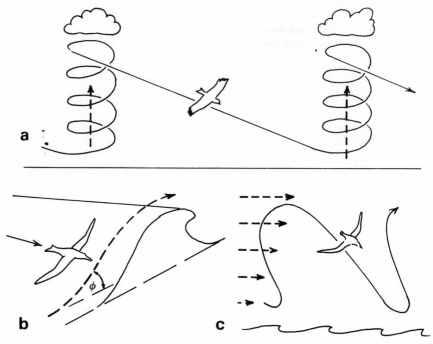

Figure 4.5 Soaring techniques. (a) Thermal soaring, (b) slope soaring over a wave and (c) dynamic soaring in a vertical wind gradient. Continuous arrows show the paths of the birds and broken arrows show the directions of air movements.

thermals by looking for likely sources on the ground, by looking for the cumulus clouds that often form at the tops of thermals and by watching soaring birds. Birds presumably use the same clues.

Successful thermal soaring requires the ability to glide in small circles (or to be more precise, in a helix of small radius). A bird flying with speed u in an arc of radius r has an acceleration u^2/r towards the centre (equation A.6). The resultant of the force mu^2/r needed to give this acceleration, and the bird's weight, must be counteracted by aerodynamic lift which must therefore be larger than either (Figure 4.4f). The lift coefficient cannot exceed some maximum value $C_{L\,\text{max}}$ so

$$mu^2/r < \tfrac{1}{2}\rho S_p u^2 C_{L\,\text{max}}$$
$$r > 2N/\rho g C_{L\,\text{max}} \qquad\qquad 4.14$$

(using equation A.17). The smaller the wing loading N $(=mg/S_p)$ the smaller the circles the bird can glide in and the smaller the thermals it can use. A typical vulture (*Gyps africanus*) would have a mass of 5 kg and a wing area of $0.7 \, m^2$, giving a wing loading of 70 Pa. The maximum lift coefficient would probably be about 1.5. Hence the minimum circling radius given by the inequality 4.14 is 8 m. This is an unattainable minimum, and *Gyps africanus* usually circles with radii between 15 and 25 m.

Since inequality 4.14 points to an advantage of low wing loading, it suggests that small gliding animals might be good thermal soarers. There are however other, partly conflicting, requirements for thermal soaring. If thermals are far apart, the animal with the lowest minimum gliding angle will be best able to glide from one to the next. There is no direct relationship between minimum gliding angle and wing loading (equation 4.12) but large gliding animals achieve lower minimum gliding angles than do small ones (Figure .4.2b). This suggests that the best thermal soarers might be large birds with unusually low wing loading for their size. The most impressive thermal soarers are vultures and storks, which fit this description.

These birds have much lower aspect ratios than some other birds such as albatrosses (Figure 4.3). Equation 4.12 shows that they would have smaller minimum gliding angles if their aspect ratios were higher, but higher aspect ratios with the same wing areas would mean even longer wings. The lift on the wings would act further from the shoulder and exert larger moments about the wing base, so the wing bones and the wing muscles would have to be stronger. The effect would be particularly severe for thermal soarers because the lift is considerably greater than body weight when they circle with a small radius.

Vultures in East Africa spend much of the day soaring, watching for dead mammals. Rüppell's griffon vulture (*Gyps rüppellii*) nests on cliffs at the edge of the Serengeti plains and travels daily by thermal soaring between the nest and the herds of mammals among which it finds its food. These herds may be up to 140 km away. Pennycuick (1972) followed commuting vultures in a motor glider. One *Gyps rüppellii* returning to its nest travelled 75 km in 96 minutes, entirely by soaring. It circled in five thermals and passed through others without circling. Storks (*Ciconia ciconia*) migrate between Europe and Africa by soaring. They travel by Suez and so make the whole journey by soaring: thermals are not found over the sea. In contrast, small passerine migrants do not soar and take a more direct route across the Mediterranean. Cranes (*Grus grus*) travel partly by flapping flight and cross the Mediterranean, but soar in thermals when opportunities occur.

Though much smaller than these birds, Monarch butterflies (*Danaus plexippus*, wing span 0.11 m) use thermal soaring in their migrations between Canada and Mexico (Gibo and Pallett, 1979). Gliding tests with dead specimens showed that their minimum sinking speed was only 0.6 m/s (so they could gain height in weak thermals) but their minimum gliding angle was 16° (so they could not glide far between thermals) and the corresponding speed was only 2.6 m/s (so they could not make headway against any but the lightest winds).

Thermal soaring is limited to daytime, in sunny weather. On the Serengeti plains, conditions for soaring are generally excellent in the dry seasons, but the thermals do not become strong enough for vultures to soar until about two hours after sunrise. In the rainy seasons, cloudy weather may prevent soaring or restrict it to a small part of the day.

Slope soaring

Many birds soar where the wind is deflected upwards by sloping ground or by waves (Figure 4.5b). Suppose the wind has velocity U, at an angle ϕ to the horizontal. A bird capable of gliding at this speed and angle can face into the wind and remain stationary relative to the ground. Kestrels (*Falco tinnunculus*) often do this when searching for prey on sloping ground. Alternatively, if the bird can glide faster than U with a sinking speed no greater than $U \sin \phi$, it can glide along the slope (Figure 4.5b). It may soar backwards and forwards to stay near one place, or continue in one direction along the whole length of the slope. Gulls (*Larus* spp.) and fulmars (*Fulmarus*) do a lot of slope soaring along coastal cliffs. The smaller petrels travel by slope soaring along ocean waves.

The requirement for slope soaring is to be able to glide at a speed of at least U with a sinking speed no greater than $U \sin \phi$. The best slope soarers should have low minimum gliding angles, at high speeds. Low gliding angles are achieved only by large gliders (Figure 4.2b) and require large aspect ratios (equation 4.12). High speeds at minimum gliding angle require large wing loadings, especially if the aspect ratio is high (equation 4.11). This makes it seem likely that the best slope soarers will be fairly large, with high aspect ratios and high wing loading for their size. Albatrosses (Figure 4.3b) fit this description. They slope soar along waves as well as using the different soaring technique that will be described in the next subsection (Wilson, 1975). Gulls and fulmars are smaller but have high aspect ratios, and fairly high wing loadings for their size. If their wing loadings were higher they might be unable to fly slowly enough to land safely on cliffs.

Monarch butterflies slope soar to climb over buildings with a following wind. Butterflies of another species (*Iphiclides podalirius*) have been observed slope soaring near a cliff, making glides without wingbeats lasting up to 30 s. One that was watched for almost an hour spent 80 % of the time gliding. The use that butterflies and other insects can make of slope soaring is limited because they cannot glide fast.

Wind-gradient soaring

Thermal soaring and slope soaring depend on vertical components of movement of the air, but another soaring technique exploits gradients of horizontal wind speed, and this is used by albatrosses.

There is a boundary layer over the sea in which the wind speed is less than it is higher up (Figure 4.5c). The shorter broken arrows in the figure represent lower wind speeds. Albatrosses glide downwind, gaining speed and so kinetic energy. Near the surface of the sea they turn into the wind and rise, converting kinetic into potential energy. Their speed relative to the ground decreases, but their speed relative to the air may not, because they are rising into faster-moving air. Consequently they can rise quite high while still keeping their airspeed high enough for the lift on their wings to support them. At a height of (typically) 12 m they turn and glide downwind again. They continue for hours, hardly ever flapping their wings.

In the discussion that follows, the kinetic energy of an albatross will be calculated relative to a moving system of coordinates. To see how this may be appropriate, imagine a man running along the corridor of a moving train and colliding with another passenger. If his kinetic energy is to be calculated for an analysis of the accident, it should be calculated from his speed relative to the train, not relative to the ground.

Consider an albatross of mass m in the climbing phase of its flight. At height h it has speed u relative to the surrounding air. In rising a small further height δh it gains potential energy $mg \cdot \delta h$ but its speed *relative to the air at height h* falls to $(u - \delta u)$, so that its kinetic energy relative to that air falls from $\frac{1}{2}mu^2$ to $\frac{1}{2}m(u - \delta u)^2$. This decrease is $\frac{1}{2}m[2u \cdot \delta u - (\delta u)^2]$, or approximately $mu \cdot \delta u$ as δu is small. If work done against drag can be neglected,

$$mu \cdot \delta u = mg \cdot \delta h$$
$$\delta u = (g/u)\delta h$$

4.15

This is the change in speed relative to the air at height h, but the bird is now at height $h + \delta h$. If its speed relative to the immediately surrounding

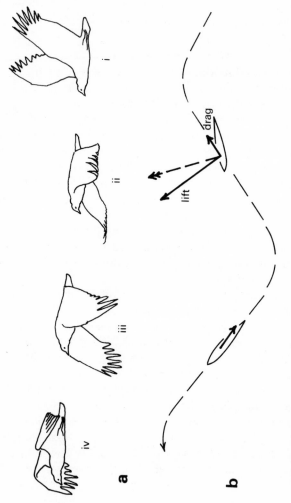

Figure 4.6 (a) Outlines traced from films of condors (*Vultur gryphus*) in flight. (From J. McGahan (1973) *J. exp. Biol.* **58**, 239–253.)
(b) A diagram representing the path through the air of a chordwise section of the condor's wing. The section, and the components of the aerodynamic force on it, are shown at two stages in the cycle of movements. (A chordwise section is one cut at right angles to the wing span.)

air is to be no less than before, the speed U of the wind relative to the ground must increase by at least δu over the same interval of height. The wind gradient dU/dh must be at least g/u.

This argument indicates that the higher the gliding speed u, the smaller the wind gradients in which soaring is possible. A more realistic argument taking account of drag would lead to the same conclusion. The high wing loadings of albatrosses are appropriate for fast gliding.

Fast flapping flight

In soaring, the work done against drag is supplied by natural air movements. In flapping flight it is supplied by the wing muscles. Flapping flight may keep an animal stationary relative to the air, in which case it is called hovering. Alternatively, it may propel the animal through the air. Slow forward flight is little different from hovering, so the distinction between forward and hovering flight is not sharp. It will be convenient to make a different distinction, between "fast" and "slow" flight. In "fast" flight the mean speed of the wing tips is little greater than that of the trunk, but in "slow" flight the wingtips move much faster than the trunk.

By this criterion, the flight of the condor shown in Figure 4.6 is fast. In each cycle of wing movements the trunk advanced 8 m and each wing top traced a curved path about 9 m long.

The wings flap up and down. In the downstroke the primary feathers (the principal feathers of the outer part of the wing) are bent upwards, showing that an upward force acts on them. In the upstroke there is no obvious bending of any of the feathers. The downstroke appears to be the power stroke and the upstroke merely the means of getting the wing into position for the next downstroke. The condor is a very large bird (males have masses of about 12 kg), but similar movements with feathers bending in the downstroke only are seen in films of birds of all sizes.

The film of the condor shows that in the downstroke, chordwise sections of the wing are roughly horizontal (Figure 4.6b). In the upstroke they are tilted slightly, leading edge up (Figure 4.6a,iv). Figure 4.6b shows that in the downstroke the wing has a positive angle of attack, so lift and drag act on it. The lift acts at right angles to the sloping path of each wing section and the resultant of lift and drag probably acts upwards and forwards. In the upstroke any angle of attack is small and the principal force on the wing is probably a small drag. The mean aerodynamic force on the two wings must have an upward component equal to the bird's weight and a forward component equal to the drag on its trunk. Compare Figure 4.6

with 2.12, which shows the swimming action of a penguin. The penguin is swimming in a medium of about the same density as itself so any upward force on the wings in their downstroke must be matched by a downward force in the upstroke.

Some large insects fly in essentially the same way as the condor. The most detailed studies have been made of locusts, suspended on a light pendulum in a wind tunnel. In these circumstances they make apparently normal flying movements. The speed of the wind was adjusted automatically so that the thrust produced by the wings balanced the drag on the trunk. The pendulum was suspended from a balance, and films were taken when the balance reading showed that the upward force produced by the wings was about equal to the weight of the locust. Thus the external forces on the locust were the same as if it had been flying through still air at a speed equal to the wind speed. Films obtained in this way show that the wings have positive angles of attack in the downstroke and small or (in part of the forewing) negative ones in the upstroke. The negative angles of attack would result in downward lift, were it not that the forewing is slightly pleated in the upstroke in a way that gives it peculiar aerodynamic properties.

The movements of locusts in this artificial situation might differ from the normal movements of free locusts. Films of wild locusts flying have been obtained by running a cine camera in the middle of locust swarms in Australia and New Guinea (Baker and Cooter, 1979). The films are not clear enough for measurements of angle of attack, but they show that the free flying action is very like the action filmed in the wind tunnel.

The balance used in the original wind-tunnel experiments registered only the mean vertical force, over many cycles of wing movements. More recent experiments with similar equipment incorporated a piezoelectric transducer capable of registering the fluctuations of force that occur in each cycle (Cloupeau, Devillers and Devezeaux, 1979). They showed that the wings produce a large upward force in the downstroke, and very little vertical force in the upstroke.

Yet other experiments with similar equipment have shown that flying locusts turn by giving their forewings different angles of attack in the downstroke. To turn to the left, the right forewing is given the larger angle of attack.

The drag D on a powered aircraft, like the drag on a glider, consists of profile drag and induced drag. It can be calculated for a fixed-wing aircraft, by using equation A.19, as

$$D = (mg/2)[(\rho C_{DO}u^2/N) + (N/\rho Au^2)]$$ 4.16

where N is the wing loading. The power P required for flight is this drag multiplied by the speed

$$P = (mg/2)[(\rho C_{DO}u^3/N) + (N/\rho Au)]$$ 4.17

Notice how similar these equations are to 4.8 and 4.9. The drag (and so the work needed to travel unit distance) is least at a particular speed, equal to

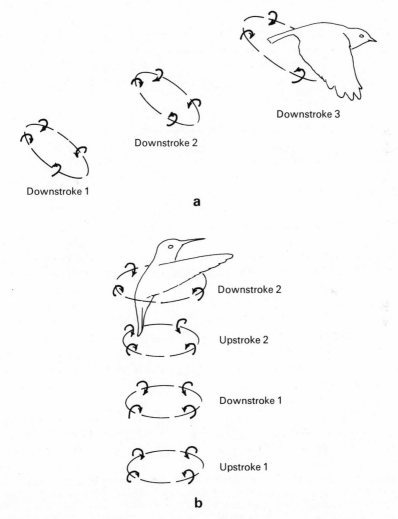

Downstroke 3

Downstroke 2

Downstroke 1

a

Downstroke 2

Upstroke 2

Downstroke 1

Upstroke 1

b

Figure 4.7 Diagrams showing air movements in the waves of birds (a) flying forward and (b) hovering. Labels "Downstroke 1" etc. show when each vortex ring was formed.

the speed u_3 (equation 4.11) that gives the minimum gliding angle when the engine is not running. The power is least at a lower speed, equal to the speed u_2 (equation 4.10) that minimizes the sinking speed in gliding.

The equations intended for fixed-wing aircraft may seem inappropriate for animals that flap their wings, but probably give reasonably good estimates of power for fast forward flight. Unsteady effects occur when a

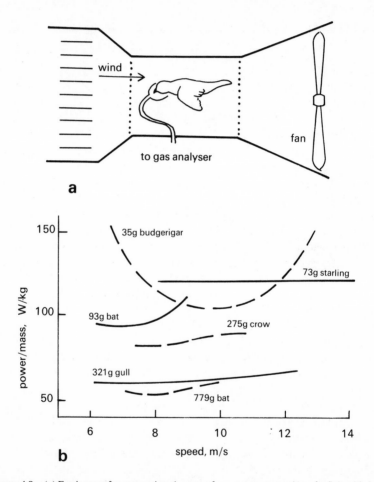

Figure 4.8 (a) Equipment for measuring the rate of oxygen consumption of a flying bird.
(b) The metabolic power (per unit body mass) used by animals flying at different speeds. Power was calculated from measurements of oxygen consumption. Data from S. P. Thomas (1975) *J. exp. Biol.* **63**, 273–294 and J. R. Torre-Bueno and J. Larochelle (1978) *J. exp. Biol.* **75**, 223–229.

wing accelerates, decelerates, or changes its angle of attack: the aero-dynamic forces on it are not the same as if it were moving steadily at the same velocity. These effects, however, are unlikely to be significant if the wing travels more than about 12 chord lengths in each wing-beat cycle. (The chord length is the distance from the leading to the trailing edge—see Figure A.5a). A condor flying fast travelled 11 chord lengths in each wing-beat cycle. Locusts travel about 25 fore-wing chords, or 12 hind-wing chords, in each cycle. Unsteady effects are unlikely to be very important in either case.

A new approach to flapping flight devised by Rayner (1979) is never-theless attractive. It has been shown that in fast flapping flight, large aerodynamic forces act on the wings only in the downstroke. Each down-stroke accelerates air downwards and backwards, producing an upward and forward force on the animal. The puff of air from each downstroke must form into a vortex ring so that the bird leaves a trail of vortex rings in its wake (Figure 4.7a). The work done against induced drag in each downstroke equals the kinetic energy of the vortex ring, which can be calculated. It has been shown that the vortex rings exist by taking photo-graphs of birds flying through clouds of dust (Kokshaysky, 1979). The rotating cores of the vortex rings appeared as dust-free regions in the photographs because they were cleared of dust by centrifugal effects.

Measurements have been made of the rates of oxygen consumption of birds and bats flying in wind tunnels (Tucker, 1968). In most cases the bird was fitted with a loose-fitting plastic mask (Figure 4.8a). Air was sucked out of the mask for analysis, and replaced by fresh air flowing in at the rear edge of the mask. All the air that the bird breathed out went to the analyser, so the rate of oxygen consumption could be calculated. In other experiments the wind tunnel was a closed system around which air circu-lated like the water tunnel of Figure 2.7a. The rate at which the oxygen concentration fell was monitored, and there was no need to attach a mask to the bird.

Some results are shown in Figure 4.8b. For budgerigars there is a speed at which the rate of oxygen consumption and the power calculated for it are least, as predicted from equation 4.17. Some other birds unexpectedly use about the same metabolic power over a wide range of speeds.

Figure 4.2b showed that gliding birds of various species lose height at minimum speeds between 1 and 2.5 m/s. They thus lose potential energy at rates g times this, i.e. between 10 and 25 W (kg body mass)$^{-1}$. If the work done against drag were supplied by muscles of efficiency 0.25 instead of being supplied from potential energy, the metabolic power requirements would be 40 to 100 W/kg. This is approximately the range of the measured powers for level flapping flight shown in Figure 4.8b. The measured powers include a little power used for purposes other than flight: the resting metabolic rates of the animals lay between 4 and 20 W/kg.

Soaring is much more economical of metabolic energy than flapping flight, as might be expected. Herring gulls (*Larus argentatus*) gliding in a wind tunnel used only 15 W/kg but laughing gulls (*L. atricilla*) used 70 W/kg in flapping flight. The resting metabolic rates were 7 and 10 W/kg, respectively.

Small birds often fly by short bursts of wing flapping, alternating with intervals in which the wings are folded against the body. For instance, the siskin (*Carduelis spinus*, mass 14 g) typically makes about 5 cycles of wing beats in a burst lasting 0.25 s, and then folds its wings for about 0.35 s. The bird rises and falls as it flies so this is called bounding flight.

Bounding flight may save energy, because drag is reduced while the wings are folded. The principle will be demonstrated by means of the equations for fixed-wing aircraft.

Consider a bird that flaps its wings for a fraction β of the time and folds them for a fraction $(1 - \beta)$. The drag on the body acts whether the wings are spread or folded and will be neglected. While they are folded, it is the only component of drag. While the wings are flapping, they have to produce lift mg/β so that the mean lift equals the weight mg of the body. Induced drag is proportional to the square of lift so equation 4.17 must be modified to obtain the power during flapping.

$$P_{\text{flap}} = (mg/2)[(\rho C'_{DO} u^3/N) + (N/\beta^2 \rho A u)] \qquad 4.18$$

(Since drag on the body is being neglected, the drag coefficient C'_{DO} refers to drag on the wings only, and is less than the coefficient C_{DO}.) This power is exerted only for a fraction β of the time so the mean power $\bar{P}$ is

$$\bar{P} = (mg/2)[(\beta \rho C'_{DO} u^3/N) + (N/\beta \rho A u)] \qquad 4.19$$

From this equation it can be shown by calculus that there is an optimum value of β which minimizes $\bar{P}$, for any particular speed. At low speeds this optimum value is greater than 1 and so is impossible: bounding flight wastes energy at low speeds. At high speeds the optimum value is less than 1 and bounding flight can save power. The critical speed at which bounding flight becomes useful is slightly greater than the speed u_3 (equation 4.11) at which drag is least in continuous flight.

To reduce mean power in this way, the bird must exert increased power during the bursts of flapping. Large birds probably cannot produce enough power in their bursts of flapping, to benefit significantly from bounding flight. The power per unit weight ($\bar{P}/mg$) required for flight at speed u_3 is $(N/\rho)^{1/2}(C_{DO}/A^3)^{1/4}$. This is greater for large birds than for small ones, because of their greater wing loadings N. However, large birds generally have about the same proportion of flight muscle in their bodies as small ones and cannot be expected to produce any more power per unit weight. Indeed, they probably produce less because their wings beat at lower frequencies.

Hovering and slow flight

This section is about flapping flight in which the animal remains stationary (hovers) or moves at a speed much less than the speed of its wingtips. Hovering is seen in hummingbirds and moths, as they drink nectar from flowers; in anisopteran dragonflies, watching for prey; and in hoverflies (Syrphinae, Diptera). Many small birds hover briefly at times, for instance if they return to a nest box and find the entrance temporarily blocked. Birds taking off from the ground, without a run, have to start with slow flight. The fastest flight of small insects, such as house flies, is "slow" flight in the sense that has been defined.

Figure 4.9 shows four techniques of hovering (Dathe and Oehme, 1978; Weis-Fogh, 1973; Savage, Newman and Wong, 1979). In three of them the body is steeply tilted. Technique (a) is used by crows and by various small passerine birds. In the down (forward) stroke the wing has a positive angle of attack and produces upward lift. In the up (backward) stroke it has little or no angle of attack and produces only small forces. Only the downstroke produces useful forces, as in fast flapping flight. Technique (b) is used by various other birds including pigeons, gulls and parrots. In the upstroke the wing has a very large angle of attack and would stall, were it not that the feathers separate and act as separate aerofoils. As in other birds, the primary feathers are asymmetrical with their shafts near the front edge of the vane. Air impinging on the upper surface of the wing twists them so that they "open" like the slats of a venetian blind. (Air impinging on the lower surface in the downstroke does not separate the feathers, because of the way they overlap.) The twisted feathers have positive angles of attack and it is clear from the way they bend (bottom picture) that lift acts upwards on them. Technique (c) is used by hummingbirds and also by many insects including moths, bees and beetles. There is no separation of feathers. The wing has a small positive angle of attack both in the down-stroke and in the upstroke. For the upstroke it turns upside-down so that

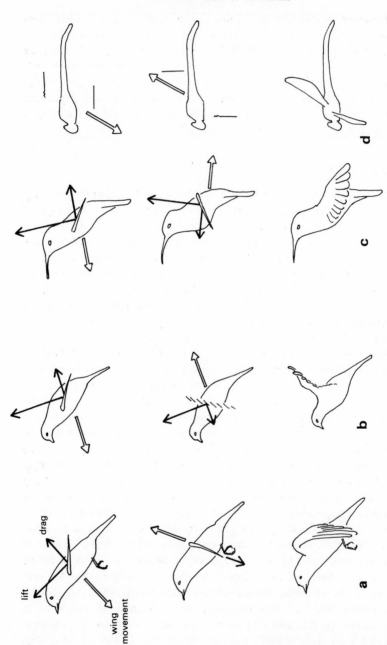

Figure 4.9 Diagrams of the hovering techniques of (a) crows and many other passerines; (b) pigeons; (c) hummingbirds; and (d) dragonflies. The upper diagrams show sections of the wings, and the forces believed to act on them during the downstroke. The middle and lower diagrams show the upstroke. Based on films from various sources.

the relatively stiff anterior edge of the wing remains the leading edge. Techniques (a) and (b) involve bending the wrist in the upstroke but (c) does not (and so is possible for insects as well as birds).

Hoverflies and dragonflies hover with their bodies horizontal, beating their wings up and down. The wings beat through small angles so that the outer parts of the wings of the dragonfly *Aeschna*, for instance, move through only four chord lengths in each stroke. The aerodynamic forces that support the insect's weight are probably produced mainly by unsteady effects. Figure 4.9d shows the hovering technique of a dragonfly. The fore- and hindwings beat almost half a cycle out of phase with each other. Much of the lift is believed to be produced by unsteady effects associated with the deceleration and rotation of the wing at the bottom of the downstroke.

The importance of unsteady effects in cases like this can be shown by calculating lift coefficients (Weis-Fogh, 1973). Different parts of each wing move at different speeds, which change in the course of a stroke. It is nevertheless possible to write equations like A.17 and A.18, giving the lift and drag that would act on a particular part of the wing at a particular stage of a stroke. The equations can be integrated to obtain an equation for the average vertical force on the complete set of wings, over a complete cycle of wing movements. Since this force must equal body weight, the equation can be used to calculate a mean lift coefficient. The procedure is analogous to a standard procedure for calculating forces on helicopter rotors. Calculations like this have given lift coefficients of 6 for dragonflies, 2–3 for hoverflies and about 3 for *Encarsia*, a tiny wasp. The highest lift coefficients that have been obtained with aerofoils in steady state experiments, in the same range of Reynolds numbers, are about 1. This shows that unsteady effects are so important that it is quite inappropriate to use equations A.17 and A.18 in these cases. (The equations are, however, appropriate for helicopter rotors which revolve continuously in one direction and quickly reach a steady state.)

The best-understood of the unsteady effects is the "clap and fling" mechanism, first explained by Professor Torkel Weis-Fogh (1973). Figure 4.10 shows how it is used by *Encarsia*. The wing movements of *Encarsia* are like those of most other hovering insects (Figure 4.9c) but the left and right wings are clapped together at the top of the upstroke (Figure 4.10a). In the downstroke, the anterior edges separate first, and air flows round them to fill the growing space between the wings (Figure 4.10b). The air movement continues after the wings have separated, as a circulation in the direction required to give upward lift (Figure 4.10c—compare Figure A.5e). The circulation is much stronger and is developed much

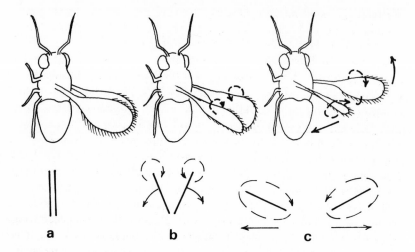

Figure 4.10   Diagrams illustrating the hovering technique of the wasp *Encarsia*. The wings are about 0.5 mm long, fringed with setae. There are two pairs but they have been drawn for simplicity as a single pair. Continuous arrows indicate wing movements and broken arrows indicate air movements. The lower diagrams are sections through the wings. Based on high-speed cine films by T. Weis-Fogh (1973) *J. exp. Biol.* **59**, 169–230.

sooner than if the wings had not started from the clapped-together position. If *Encarsia* did not have the benefit of unsteady effects it would have to beat its wings at a higher frequency.

Many other animals use the clap and fling in slow flight. Some films of climbing flight of locusts show the hindwings clapping together, and the wings of pigeons can be heard clapping together at the top of the upstroke, when they take off from the ground.

The wingbeats of hovering animals drive puffs of air downwards, as vortex rings (Ellington, 1978; Rayner, 1979). If the wing-beat frequency is high enough, the vortex rings formed by successive beats merge, forming a downward jet of air like the jet below a helicopter rotor (Figure 4.7b). Calculations show that the rings must merge in most insects but remain separate in many birds, especially in those that produce lift on the down-stroke only and so form only one vortex ring in each wing-beat cycle (Figure 4.9a—the other mechanisms form two rings per cycle).

For the animals whose vortex rings merge, the induced power for flight can be calculated from helicopter theory. More power is needed if the rings do not merge. The induced power P_i required for hovering by a helicopter of weight mg with a rotor of radius r is

$$P_i = (m^3 g^3 / 2\rho\pi r^2)^{1/2} \qquad\qquad 4.20$$

where ρ is the density of the air (equation A.20). For animals, r is approximately the length of the wings. If hovering animals of different sizes were geometrically similar, r would be proportional to $m^{1/3}$ and P_i would be proportional to $m^{7/6}$. Large animals would require more power to hover, per unit body mass, than small ones. It is difficult for large animals to produce enough power for hovering. The largest animals able to maintain oxygen balance while hovering seem to be large hummingbirds of mass about 20 g. Birds up to about the size of domestic pigeons (0.4 kg) can hover briefly and large birds cannot hover at all.

Now consider the frequency of wing movements. Too low a frequency results in separate vortex rings instead of a continuous jet, and increases the induced power above the value given by helicopter theory. Too high a frequency involves moving the wings unnecessarily fast and increases the power expended against profile drag. It seems likely that the optimum frequencies, for animals of different size, will produce geometrically similar wakes, i.e. wakes in which the distance between vortex rings is proportional to r. Let the downward velocity of the air passing the wings be v, and let the wing-beat frequency be f. Each cycle of wing movements produces two vortex rings so the distance between successive rings is $v/2f$. To make this proportional to r,

$$f \propto v/r$$

The theory of helicopters from which equation 4.20 is derived shows $v \propto m^{1/2}/r$, so the condition is

$$f \propto m^{1/2}/r^2 \qquad\qquad 4.21$$

For geometrically similar animals, $r \propto m^{1/3}$ and so $f \propto m^{-1/6}$. Small animals would beat their wings at higher frequencies than large ones, as they are observed to do.

Hovering animals of different sizes are not geometrically similar. They tend to have wing lengths about proportional to $m^{0.4}$ and frequencies about proportional to $m^{-0.3}$. These proportionalities satisfy condition 4.21. Also, $r \propto m^{0.4}$ makes the induced power P_i proportional to $m^{1.1}$ instead of $m^{7/6}$ ($= m^{1.17}$), so that the difficulty of hovering does not increase as rapidly with increasing size as it would for geometrically similar animals.

Induced power is the main aerodynamic power requirement for large hovering animals, especially for those that produce only one vortex ring in each cycle of wing beats. For small insects, power expended against profile drag is more important because their wings work at low Reynolds numbers.

Yet another power requirement is liable to arise. The wings have to be given kinetic energy as they are accelerated for each stroke. If this kinetic energy were small compared to the aerodynamic work done in the stroke, it could be converted to aerodynamic work at the end of the stroke by letting drag bring the wings to rest. It has been calculated, however, that it is as large or larger than the aerodynamic work, for many hovering animals (Weis-Fogh, 1973). The wings must therefore be brought to rest largely by other means, for instance by muscles doing negative work. If they were brought to rest in this way the kinetic energy would be degraded to heat and lost. The wing muscles would have to do work in the next stroke to replace this loss, and this would be a major part of the energy cost of flight.

This cost can be avoided if the wings are mounted on elastic structures, so that they vibrate with a natural frequency equal to the wing-beat frequency. In such a system, kinetic energy is converted to elastic strain energy at the end of each stroke, and back again in an elastic recoil at the beginning of the next stroke. The same principle is illustrated by a guitar string that continues vibrating long after being plucked, as energy is converted back and forth between the kinetic and elastic forms.

The flight muscles themselves have elastic properties. The strain energy they can store is about 1–2 J/kg, while the power output of the flight muscles of insects of all sizes seems to be about 200 W/kg, close to the maximum of which muscle is capable. The work done by the muscles in a wing-beat cycle would thus be about 10 J/kg for a large insect with a wing-beat frequency of 20 Hz and 1 J/kg for a small one with a frequency of 200 Hz. Muscle elasticity can make useful savings of energy for small insects but not for large ones.

Large insects such as dragonflies and locusts have supplementary elastic energy stores made of the protein resilin. This protein has elastic properties very like those of soft rubber. In dragonflies, it is found in the apodemes (tendons) of the wing muscles, and in locusts it forms flexible elastic hinges joining the wings to the body.

Dipteran flies use high wingbeat frequencies, so the strain energy stored by the muscles is probably greater than the kinetic energy of the moving wings. Their thoraxes have a click mechanism that seems to take up the excess strain energy in mid-stroke. This works like an electric light switch, which is stable only in the fully up and fully down positions and has strain energy stored in a compressed spring at intermediate positions. It compensates for the excessive elastic compliance of the muscles, by storing most strain energy when they store least.

Birds have relatively low wingbeat frequencies so muscle elasticity cannot save them much energy. They do not seem to have any supplementary elastic system capable of carrying kinetic energy over from one stroke to the next, but the elasticity of the feathers may assist conversion of kinetic energy to aerodynamic work at the end of each downstroke (Pennycuick and Lock, 1976).

Take-off and landing

Animals that can hover take off from level ground by jumping into the air, hovering and gradually building up speed. The house fly, *Musca*, locusts and birds up to at least the size of pigeons take off by jumping and hovering. Birds too large to hover cannot do this. If the wind is faster than their minimum gliding speed they can take off by spreading their wings and facing into the wind, but if not they must use some other method to gain enough airspeed to start flying. Birds perched on cliffs or branches can get the necessary speed by dropping from the perch, but take-off from a flat surface requires a run. One bird which has to run to take off is the Kori bustard (*Ardeotis kori*), one of the largest flying birds, weighing up to 16 kg at least, which spends most of its time on the ground on the East African plains. Swans and many other large water birds run along the

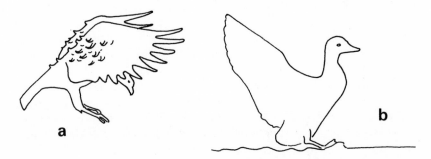

Figure 4.11 Birds landing. (a) A condor (*Vultur gryphus*) and (b) a mallard (*Anas platyrhynchos*). (a) is an outline from a film (J. McGahan (1973) *J. exp. Biol.* **58**, 239–253); (b) is based on a photograph (G. Rüppell (1977) *Bird Flight*, Van Nostrand Reinhold, New York).

surface of water, beating their wings, to reach a speed high enough for take-off.

Just as getting up speed can be a problem in take-off, slowing down before hitting the ground can be a problem in landing. Small birds can decelerate to hovering flight for a gentle landing but large birds cannot do so. Figure 4.11a shows a condor landing. It is beating its wings in a nearly horizontal plane, with so large an angle of attack on the downstroke that they have stalled. The irregular air movements over the wing, due to stalling, are making the small feathers flutter. Because they are stalled the wings must be producing less than maximum lift, but high drag. The lift may be less than the weight of the animal but if the drag is large enough the bird decelerates rapidly and is moving very slowly when it alights. Photographs of other moderate to large birds landing also show stalled wings.

A bird landing on a branch or cliff can use gravity as a braking force by swooping lower than its perch. To get the greatest possible advantage from the manoeuvre the bird should be flying as slowly as possible at the bottom of its swoop. Ducks and many other water birds land at high speed on water, but use their feet as brakes (Figure 4.11b).

CHAPTER FIVE

WALKING, RUNNING AND JUMPING

Horses walk to go slowly, trot to go faster, then canter and finally gallop at their highest speeds. The walk, trot, canter and gallop are different gaits, involving different patterns of leg movement. This chapter is about gaits and their relative merits. Most of the examples are vertebrates, but it includes a little information about insects and other arthropods.

Clearly, a method for describing gaits is needed. In a regular gait, each foot is set down just once in each stride and successive strides are identical. The *stride frequency* is the number of strides in unit time, and the *stride length* is the distance travelled in a stride (the distance from a footprint to the next print made by the same foot).

The *duty factor* of a foot is the fraction of the time for which that foot is on the ground. For instance, a walking man may have a duty factor of 0.6 for each foot, which implies that for a fraction 0.2 of the stride, both feet are on the ground simultaneously. A running man may have a duty factor of 0.3 for each foot, which implies that for 0.4 of the time both feet are off the ground simultaneously. Gaits with duty factors greater than 0.5 are generally called walks, and ones with smaller duty factors are called runs. Trots, canters and gallops are all running gaits.

The *relative phase* of a foot is the time at which it is set down, expressed as a fraction of the stride. The stride is usually reckoned to start when one particular foot is set down, so that the relative phase of that foot is zero. The relative phases of the two feet of a walking or running man are thus 0 and 0.5, but the relative phases of the feet of a hopping kangaroo are both zero.

Various gaits will be discussed in order of increasing speed, but before that two pieces of equipment must be explained. A *force platform* is a plate mounted on transducers which give electrical signals proportional to any force that acts on the plate. Force platforms are often set into the floor so that records can be made of the forces exerted by the feet of animals running over them. The most useful force platforms give separate electrical outputs signalling vertical, longitudinal and transverse components of force. Very careful design is needed, if rapidly-changing forces are to be recorded accurately.

Treadmills are simply conveyor belts, which are used in research on running in the same way as water-tunnels and wind-tunnels are used in research on swimming and flight. The animal is trained to run on the moving belt so as to remain stationary relative to the laboratory. This makes it possible to keep it running steadily at a chosen speed. Also, apparatus can be connected to the animal. For instance, the animal may be connected through a gas mask to gas analysis equipment, so that its expired air can be analysed and the rate at which it is using energy calculated. This experiment has been performed with large antelopes and young lions, as well as with more tractable animals.

Walking slowly

A wheeled vehicle moving at constant velocity over level ground is perpetually in equilibrium. The resultant of the forces on its wheels is a vertical force equal to its weight, acting through its centre of mass. It moves steadily without bouncing up and down, rolling from side to side or making any other unwanted movements.

Wheels stay in contact with the ground but legs have to be lifted and set down. Whether or not equilibrium is maintained depends on the duty factors and relative phases of the feet and on the patterns of force they exert on the ground. A running animal is obviously not in equilibrium while its feet are all off the ground. It has no need to maintain perpetual equilibrium: it is sufficient for the mean resultant force on the feet, over a complete stride, to match its weight. Any unwanted displacement due to a departure from equilibrium is quickly corrected when another foot is set down. The situation is quite different for an animal walking really slowly, like a tortoise. For it, there is so much time between footfalls that it is essential to keep very nearly in equilibrium throughout the stride.

An animal with large feet (for instance a human or a bird) can be in stable equilibrium standing on just one foot, which must be vertically under its centre of mass. An animal with small feet needs at least three on the ground. (Three legs is the minimum for a stable stool.) A vertical line through its centre of mass must pass through the triangle formed by the

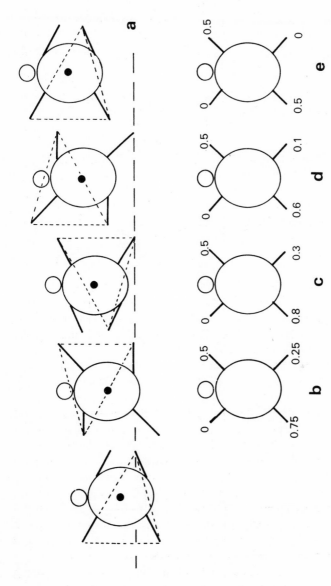

Figure 5.1 (a) Diagram showing successive positions of a quadruped walking so as to be in equilibrium at all times. In each diagram the three feet connected by a triangle are on the ground, and the fourth has just been lifted. The black dot marks the centre of mass.
(b) to (e) Diagrams showing relative phases in gaits described in the text.

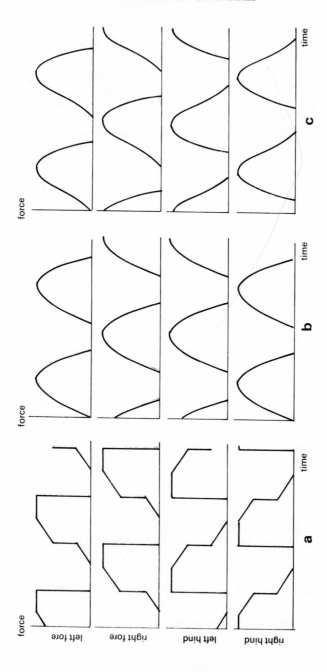

Figure 5.2 Graphs of forces on the feet, against time, for three quadrupedal gaits. The pattern of forces shown in (a) maintains equilibrium throughout the stride. Pattern (b) minimizes unwanted displacements for an animal with very slow muscles (equation 5.1) and pattern (c) for slightly faster muscles (equation 5.2).

feet (the triangle of support.) Similarly a quadruped can maintain perpetual equilibrium as it walks, if it moves its feet one at a time so that it always has three on the ground. This means that the average of the duty factors of it feet must be at least 0.75. Further, it must move its feet in such an order that its centre of mass is always over the triangle of support. It can be shown by drawing diagrams like Figure 5.1a that the only possible order, for duty factors between 0.75 and 0.83, is the one shown in the figure. (Another order is also possible at even higher duty factors.) If the duty factor is 0.75 for each foot, the feet must move with the relative phases shown in Figure 5.1b. With higher duty factors, slightly different relative phases become possible. For instance, with a duty factor of 0.8 the gait shown in Figure 5.1c is possible, and arguably better than Figure 5.1b because it allows a greater margin of stability.

Tortoises use duty factors of about 0.8 and should therefore use gaits like Figure 5.1b or (better) c, but they actually use gaits like Figure 5.1d, which means that at times there are only two feet on the ground. They cannot maintain perpetual equilibrium, walking like this. They rise and fall, pitch and roll as they walk. Why do tortoises not use a different gait and keep themselves in equilibrium (Jayes and Alexander, 1980; Alexander, 1981)?

The answer seems to be that their muscles are too slow. For equilibrium, feet must exert the right forces as well as being in appropriate positions. With duty factors of 0.75, the forces must be as in Figure 5.2a. (This can be shown by solving the equations for equilibrium for each position shown in Figure 5.1a, and intermediate positions.) With higher duty factors some variation is possible, at the stages when all four feet are on the ground together, but the abrupt changes of force shown in Figure 5.2a are still required. The forces on the feet must rise and fall instantaneously between high and low values, at certain stages of the stride.

Tortoises have very slow muscles. Since they feed on plants, they have no need to pursue prey, and since they can retire into their shells they do not need to run away from predators. Slow muscles can maintain tension at less energy cost than fast ones, because their cross-bridges need reactivating less often. (It has also been shown that tortoise muscles perform work with unusually high efficiency, and this may be due to their slowness.)

Slow muscles cannot exert forces which rise and fall abruptly, as in Figure 5.2a. Suppose that the slowness of tortoise muscle prohibits any pattern of force except a rather slow rise to a peak followed by an equally slow decline (Figure 5.3a). If that were the case, what gait would be best?

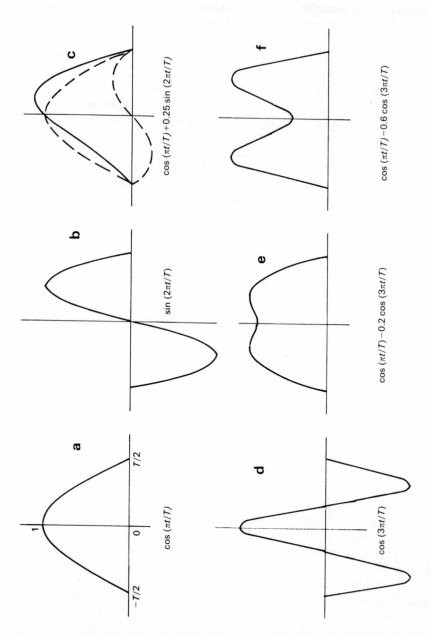

Figure 5.3 Graphs of various trigonometric functions against t, over the range $t = -T/2$ to $t = +T/2$. These graphs are referred to in discussions of the patterns of force exerted by feet on the ground.

Before the question can be answered, this pattern of force must be described mathematically, using a method that will be useful again when human walking is discussed. Let a foot be on the ground from time $-T/2$ to time $T/2$. At any time t while it is on the ground let it exert a vertical force given by

$$F = A\cos(\pi t/T) \qquad 5.1$$

where A is a constant. A mathematical model of a tortoise has been devised, in which the feet exert forces like this. Their duty factors and relative phases can be varied. The model was used to calculate for a wide range of possible gaits

(i) the range of height through which the centre of mass would rise and fall during a stride,
(ii) the range of angles through which the shell would pitch (tilt nose up or down), and
(iii) the range of angles through which it would roll (tilt from side to side).

It was found that at realistically low stride frequencies, only a very narrow range of gaits was possible. All other gaits involved such large unwanted movements of types (i), (ii) or (iii) that the shell would hit the ground at some stage in the stride. The best possible gait was the one shown in Figures 5.1e and 5.2b. It is very different from the gaits that allow equilibrium, with sufficiently fast muscles (Figure 5.1b,c), but only a little different from the gaits that tortoises use (Figure 5.1d).

Even better agreement between tortoises and mathematics was obtained by refining the theory. Suppose the muscles were just a little faster so that the feet could exert forces like

$$F = A[\cos(\pi t/T) + r\sin(2\pi t/T)] \qquad 5.2$$

where A and r are both constants. By adding the cosine and sine components (Figure 5.3a and b) a skewed pattern of force is obtained (Figure 5.3c). The model showed that one particular combination of degree of skewness (value of r) and relative phases was better than any other, and better than any gait based on equation 5.1. It is shown in Figure 5.2c. The relative phases match the gaits that tortoises use (Figure 5.1d). Records of the forces exerted by tortoise feet, obtained by letting a tortoise walk across a force platform, are very like Figure 5.2c, with similar skewing.

Tortoises have to use a particular gait because they walk so slowly.

Quadrupedal mammals do not have to, in their usual range of speeds, but some use rather similar relative phases. The difference can be shown by a simple calculation. Consider an animal with legs of height l. If it were suddenly left unsupported it would fall a distance $gt^2/2$ in time t (equation A.5). It would fall a distance l and hit the ground in time $\sqrt{2l/g}$. If the stride frequency is n, this is a fraction $n\sqrt{2l/g}$ of the duration of a stride. For a tortoise with legs 5 cm high walking with a stride frequency of 0.5 Hz, it is 0.05 stride periods. For a dog with legs 0.5 m high walking very slowly with a stride frequency of 1.2 Hz, it is 0.4 stride periods. For the same dog galloping fast with a stride frequency of 3 Hz it is 1.0 stride periods. Departures from equilibrium, lasting the same fraction of a stride, might be intolerable to a tortoise but unimportant to a dog.

Walking faster

As a person walks faster and faster he reaches a speed at which suddenly he breaks into a run. Adults start running at about 2.5 m/s, and children at lower speeds. Why do we change our gait?

A partial explanation is given by a very crude model of walking, based on the observation that people walk with their legs nearly straight. Imagine a person whose legs remain perfectly straight, while the feet are on the ground. His head and trunk bob up and down as he walks, as the bodies of real people do. They move in a series of arcs of radius l, where l is the length of a leg (Figure 5.4a). Let the speed of walking be u. In position (ii) the trunk is moving with speed u along an arc of radius l so it has an acceleration u^2/l towards the centre of the circle; that is, downwards towards the foot (equation A.6). If the mass of the legs is small enough to be ignored, in comparison with the mass of the trunk, the centre of mass of the body also has a downward acceleration u^2/l. Since the foot is not glued to the ground the man cannot pull himself downwards, but can only allow himself to fall under gravity. Consequently the downward acceleration of his centre of mass cannot be greater than the gravitational acceleration g.

$$u^2/l \leqslant g$$
$$u \leqslant \sqrt{gl} \qquad\qquad 5.3$$

Adult men have legs about 0.9 m long, and g is 9.8 m/s². Hence the maximum possible speed of walking with straight legs is $\sqrt{9.8 \times 0.9} = 3$ m/s, only a little higher than the speed at which men start running. Athletes in

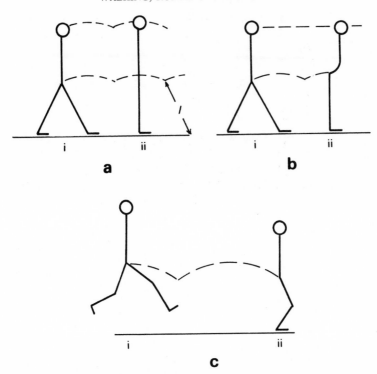

Figure 5.4 Diagrams of a person (a) walking, (b) doing a racing walk and (c) running.

walking races go faster, at about 4 m/s, but only by using a special technique (Figure 5.4b). They keep their legs fairly straight but bend the lower part of the back in position (ii), sticking their bottoms out. Their hips rise and fall in near-circular arcs but this bending makes the path of the upper parts of the body much more level. The accelerations of the centre of mass are therefore smaller than would be required for ordinary walking at the same speed. Equation 5.3 also helps to explain the peculiar hopping gait used by astronauts on the moon. There g is only 1.6 m/s^2, so the maximum speed for straight-legged walking would be 1.2 m/s, which is inconveniently slow.

Equation 5.3 can be expressed by saying that at the fastest possible speed for walking, the Froude number u^2/gl (see appendix) is 1. Froude numbers are important in studies of walking and running because these activities involve interaction between gravity and inertia.

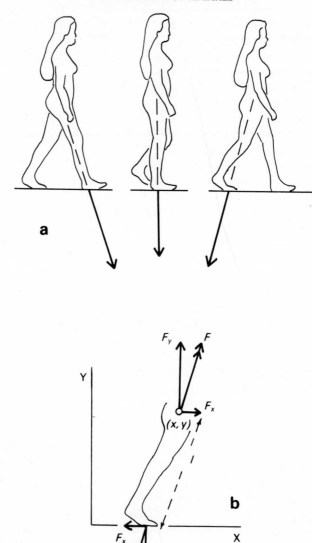

Figure 5.5   (a) Forces exerted by the right foot of a walking woman, at three stages of a step. (b) Diagram to show how the work done by a leg is calculated.

Force platform records show that tortoises exert roughly vertical forces with their feet, but people exert sloping forces at the beginning and end of each step (Figure 5.5a). Throughout the step, the force is more or less in line with the centre of mass of the body, which is a little above the hips. The decelerating effect of the forward push early in the step is compensated by the accelerating effect of the backward push later on.

Force platform records of a man walking and running are shown in Figure 5.6. In every case the records of the horizontal component show a forward force followed by a backward one. The three records of the vertical component are, however, quite different. In slow walking (Figure 5.6a) the vertical component shows two peaks with a shallow dip between them. In faster walking (Figure 5.6b) the dip becomes a deep valley. In running (Figure 5.6c) there is only one major peak (ignoring a small subsidiary peak at the start of the record, due to vibrations in the leg following impact with the ground). Why does the man change these patterns of force, as he changes his speed?

It has been argued that the maintenance of equilibrium is far less critical for mammals (including man) than it is for tortoises. It seems reasonable to guess that people adjust their gaits to minimize the energy cost of walking. How does this cost arise? A person walking at constant speed on level ground has the same potential and kinetic energy at corresponding stages of successive strides, and so is doing no net work giving himself potential or kinetic energy. He does very little work against the drag of the air (unless he is struggling against a strong wind) and very little

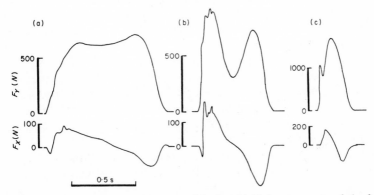

Figure 5.6 Records of the vertical (F_Y) and horizontal (F_X) components of the force exerted by the foot of a man (a) walking slowly at 0.9 m/s, (b) walking fast at 1.9 m/s and (c) running at 3.7 m/s across a force platform. From R. McN. Alexander and A. S. Jayes (1978) *J. Zool., Lond.* **185**, 27–40.

against friction in his joints. The quite substantial energy cost of walking arises because the muscles of each leg do positive work (tending to increase potential and kinetic energy) at some stages of each step and negative work (tending to decrease potential and kinetic energy) at others (Alexander, 1980, 1981). Both positive and negative muscular work require metabolic energy.

Assume (realistically) that the walker shown in Figure 5.5b does so little work against aerodynamic drag and joint friction that these can be ignored. Assume also (less realistically, but for mathematical convenience) that the mass of the whole body is concentrated in the trunk, so that the mass of the legs can be ignored. Then the components F_x and F_y of the force the leg exerts on the trunk are equal and opposite to the components of the force on the ground. At time t the coordinates of the hip are x, y and the components of the velocity of the hip are dx/dt, dy/dt. It follows from the definition of work (W, equation A.9) that the mechanical power output of the leg, dW/dt, is given by

$$dW/dt = F_x \cdot dx/dt + F_y \cdot dy/dt \qquad 5.4$$

If the resultant force F acts in line with the leg, the equation can be simplified

$$dW/dt = F \cdot dl/dt \qquad 5.5$$

where l is the length of the leg at time t. Stiff-legged walking (as shown in Figure 5.4a) is very economical of energy, in the range of speeds at which it is possible, because dl/dt is zero for most of the step. (There must be a brief stage when both feet are on the ground and dl/dt is not zero: otherwise, F would have to rise for an instant to infinity to give the body the necessary acceleration.)

At some stages of the step, dW/dt is positive, and at other stages it is negative. If the time courses of F_x and F_y have been defined, equation 5.4 can be integrated over a complete stride (adding up positive and negative work separately) to determine the work needed for a stride. Hence the metabolic energy cost of walking can be estimated, for any specified pattern of F_x and F_y.

Let a foot be on the ground from time $-T/2$ to time $T/2$. Let the vertical component F_y of the force it exerts, while on the ground, be given by

$$F_y = A[\cos(\pi t/T) - q\cos(3\pi t/T)] \qquad 5.6$$

where A and q are constants. The constant A must have whatever value is

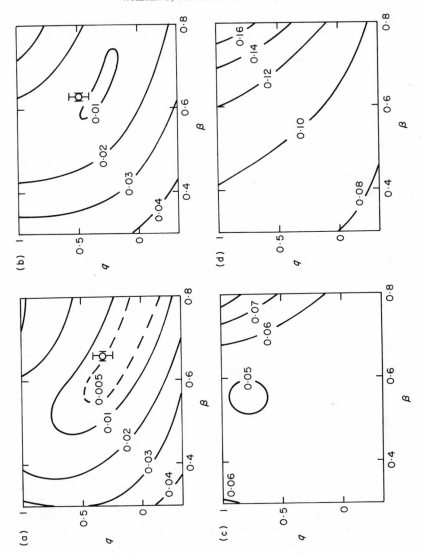

Figure 5.7 Graphs showing calculated metabolic power requirements for walking and running with different duty factors (β) and shape factors (q), at four different speeds. The contours on each graph show power requirements in arbitrary units. The speeds represented, for an adult man of average stature, are (a) 0.9 m/s, (b) 1.8 m/s, (c) 3.5 m/s and (d) 5.0 m/s. The points in (a) and (b) show the mean values of β and q used by men at the speeds in question. From R. McN. Alexander (1980) *J. Zool., Lond.* **192**, 97–117.

needed to make the mean vertical force on the ground equal to body weight. The other constant q (called the shape factor) can be given different values to simulate a wide range of different patterns of force. Figure 5.3e shows that equation 5.6 with $q = 0.2$ gives a very reasonable imitation of the force record of slow walking (Figure 5.6a). Similarly Figure 5.3f with $q = 0.6$ imitates the record of fast walking (Figure 5.6b) and Figure 5.3a with $q = 0$ imitates the record of running (Figure 5.6c). Notice that increasingly positive values of q give increasingly two-humped graphs of F_y against time. (Negative values of q give bell-shaped graphs.)

Let F_x change in the course of a step so that the resultant of F_x and F_y is always in line with some specified point in the trunk. Figure 5.6a shows that this is realistic.

These ideas are the basis of a mathematical theory of walking with three variables; the shape factor q, the duty factor β and speed (Alexander, 1980). The shape factor can only have values between -0.33 and $+1.00$. (Values outside this range give an upward force on the ground at some stage of the step.) The duty factor could in principle have any value between 0 and 1 but very low values would require very large forces and very high values would leave too little time to swing the foot forward for the next step. Figure 5.7 shows calculated power requirements for walking and running at four different speeds and for all possible shape factors and a wide range of duty factors.

The most economical way of travelling at each speed is shown by the contours of power requirement. At the lowest speed (0.9 m/s: Figure 5.7a) the contours represent a narrow valley on the right-hand side of the graph. Least power is needed at the bottom of this valley, with large duty factors (β) and with low shape factors (q). At 1.8 m/s (Figure 5.7b) there is a more clearly defined optimum at $\beta = 0.65$, $q = 0.3$. At 3.5 m/s (Figure 5.7c) the optimum has moved to $\beta = 0.55$, $q = 0.8$. All these gaits are walks since the values of β are greater than 0.5. At 4.0 m/s (not illustrated) the optimum jumps suddenly to a run and at all higher speeds (Figure 5.7d) the most economical gait is a run with low β and low q. (The theory indicates that β should be infinitesimally small but that would be impossible as it would involve infinite forces on the feet). From this it can be predicted that people should walk with increasing values of q, as they increase speed, but at a critical speed they should switch suddenly to a run, which is exactly what they do. The values of q and β that men use at 0.9 m/s and 1.8 m/s are close to the values that the theory suggests. Adult men run at speeds above about 2.5 m/s even though the theory suggests they should continue walking until they reach 4 m/s. This discrepancy seems to be due to the theory being too simple: it ignores energy savings by elastic storage which are important in running (as will be shown) but insignificant in walking with high values of q. Notice that the minimum in Figure 5.7c is so shallow that a very small change would shift the balance of advantage to running.

The energy cost of walking has often been estimated by considering the fluctuations of potential and kinetic energy that occur in each stride (Cavagna, Heglund and Taylor, 1977). The muscles must do positive work whenever the total (potential + kinetic) energy of the body increases, and negative work whenever the total decreases. This approach is inferior to the one just described because it fails to take account of the possibility that when both feet are on the ground, one leg may be doing positive work and the other negative work. This failure is particularly serious in discussions of quadrupedal walking gaits, in which there are always at least two feet on the ground.

The theory described in this section indicates that the optimum force patterns for walking quadrupeds are very like the optimum force patterns for bipeds. Nevertheless, force platform records of walking dogs, sheep and horses seldom or never show values of q greater than 0.4—I am unable to explain why these mammals do not use higher values of q, when they walk fast.

Insects very seldom use duty factors less than 0.5, so even their fast gaits are generally walks (Delcomyn, 1971). The relative phases are generally about 0, 0.5, 0 for the left feet and 0.5, 0, 0.5 for the right ones. The feet thus form two groups of three which move alternately, so that there are always at least three feet on the ground. Cockroaches (*Periplaneta*) travel at speeds ranging from 0.03 to 0.8 m/s, and use these relative phases at all but the very lowest speeds. Since the duty factor is always 0.5 or more, very long strides are impossible and high speeds require high stride frequencies, up to 24 Hz at 0.8 m/s. (This is about the same as the frequency of the wing beat in flight.)

Periplaneta walking very slowly use stride frequencies down to about 2 Hz. They carry their bodies about 1 cm from the ground so the parameter $n\sqrt{2l/g}$ (p. 88) has the low value of 0.1. This means it is probably essential for the animal to have at least three feet on the ground almost all the time, with the centre of mass over the triangle of support. At high speeds this is not the case, but it may nevertheless be desirable to keep three feet on the ground to prevent displacement by gusts of wind. Wind has very little effect on the motion of larger animals such as mammals because their ratios of surface area to mass are much smaller.

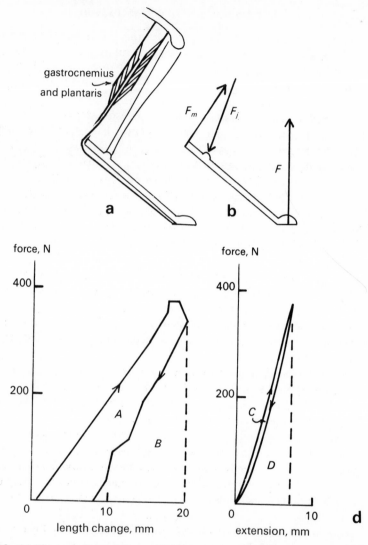

Figure 5.8 (a) Diagram of the skeleton of a kangaroo's leg, with the extensor muscles of the ankle.

(b) Diagram showing forces on a kangaroo's foot.

(c) Graph of force against length for the plantaris muscle and tendon of a wallaby (*Protemnodon*) showing how they extended and then shortened while the foot was on the ground during hopping at 2.4 m/s.

(d) A graph of force against length for the plantaris tendon alone, corresponding to (c).

(c) and (d) are based on data of R. McN. Alexander and A. Vernon (1975) *J. Zool., Lond.* **177**, 265–303. It has been assumed in (d) that the properties of the tendon resemble those of sheep plantaris tendon, as measured by R. F. Ker (1981) *J. exp. Biol.* **93**, 283–302.

Running bipeds

While a running man has his feet off the ground he rises and then falls under gravity. The man shown in Figure 5.4c, position (i), is at the top of his trajectory. He is falling when his foot hits the ground and must bend his leg. If the shape factor has a low value (as it has in all force platform records of running that have been made, of any species), the leg is most bent in the middle of the step, in position (ii). Thus dl/dt (equation 5.5) is first negative and then positive: the leg does negative work followed by positive work.

A spring does negative work (storing elastic strain energy) as the force on it increases, and positive work (in the elastic recoil) as the force decreases. The force on a running man's foot rises from zero to a maximum, in position (ii), and then falls. Thus the function of the leg muscles in a running man could in principle be taken over by a spring made of some elastic material: he could bounce along like a rubber ball (Cavagna, Heglund and Taylor, 1977).

To a large extent this happens. The most important of the elastic structures involved are tendons, especially those of the extensor muscles of the ankle joint. The evidence for this has been worked out in more detail for hopping kangaroos than for running men (Alexander and Vernon, 1975). All the statements in the past few paragraphs are as true of hopping as of running.

Figure 5.8a is a diagram of a kangaroo leg showing the extensor muscles of the ankle, the gastrocnemius and plantaris. Both are pennate with long tendons and relatively short muscle fibres. The tendon of the gastrocnemius inserts on the heel and that of the plantaris runs round the heel to the toes, but both exert moments about the ankle. At the instant shown in Figure 5.8b, the ground is exerting a force F on the foot. The moment of F about the ankle must be balanced by the contrary moment of the force F_m exerted by the muscles. (The weight of the foot and the inertia force due to its acceleration are small enough to be ignored in this context. F_j is the reaction at the ankle joint.) Thus if the size and direction of F are known from a force platform record and the position of the leg is known from a film taken at the same time, F_m can be calculated. Also, if the angles of the joints are measured from successive frames of the film the changes in length of the muscles (with their tendons) can be calculated.

Figure 5.8c shows data obtained in this way for a wallaby. It has been assumed that the force F_m, exerted by the gastrocnemius and plantaris muscles together, is shared between them in such a way that the mean stress in the muscle fibres is the same for both muscles. Since Figure 5.8c is a graph of force against length, areas under the graph represent work. As the plantaris muscle is stretched it does negative work $-(A+B)$. As it shortens again it does positive work B. (Though it does more negative than positive work other leg muscles do more positive than negative work.)

Some of the extension and shortening shown in Figure 5.8c is due to changes in length of the muscle fibres, and some to elastic extension and recoil of the tendon. The part due to the tendon has been estimated from the force, the dimensions of the tendon and the properties of sheep plantaris tendon (Ker, 1981) as the properties of wallaby tendon have not been measured. Figure 5.8d shows that the tendon does negative work $-(C+D)$ as it is stretched and positive work D as it recoils. These are quite large fractions of the work $-(A+B)$ and B that would otherwise have to be done by the contractile apparatus of the muscle fibres, so useful savings of metabolic energy are made. Similar savings must be made by the gastrocnemius tendon. Further savings are made by the elastic behaviour of the muscle fibres themselves but it has been shown by experiment that these savings are small, compared to the savings due to the tendons (Morgan, Proske and Warren, 1978).

Such calculations suggest that the elastic properties of the gastrocnemius and plantaris muscles and tendons save one third or more of the metabolic energy that would otherwise be required for hopping. They are based on the assumption of equal stresses in the two muscles. More energy would be saved if one of the muscles exerted all the force, because a given force causes more extension if it is applied to one tendon than if it is shared between two. Also, bigger percentage savings would be made at higher speeds. As they increase speeds, wallabies and other mammals lengthen their strides and reduce their duty factors. The feet are on the ground for a smaller fraction of the time, at high speeds, so the forces they have to exert while on the ground are larger. Though the forces are greater and stretch the tendons more, the changes in length of the leg are about the same as at lower speeds. Larger fractions of the positive and negative work (calculated from $F \cdot \delta l$, equation 5.5) are provided by the elastic behaviour of the tendons.

Measurements have been made of the oxygen consumption of kangaroos hopping on a treadmill at speeds up to 6 m/s. At all speeds of hopping an 18 kg kangaroo used about 20 cm^3 oxygen/s, corresponding to a metabolic power consumption of 400 W. This is about twenty times the resting rate of oxygen consumption, so about 380 W was presumably being used for hopping. Kangaroos were also made to hop across a force platform and the force records were used to calculate the work done by the leg muscles in each stride. Multiplying the work by the stride frequency showed that an 18 kg kangaroo hopping at 6 m/s does positive and negative work at rates equal to +190 W and −190 W. Human muscles (and presumably kangaroo muscles) do positive and negative work with efficiencies 0.25 and −1.2 (p. 4). The kangaroo might therefore be expected to use metabolic energy at a rate $(190/0.25)+(-190/-1.2) = 918$ W. It actually only used 380 W, so 59% of the energy that would otherwise have been needed for hopping was apparently being saved by elastic storage. Savings calculated in the same way for lower speeds were lower.

Energy is also saved by elastic storage in the running gaits of other large mammals such as people, dogs and antelopes, and also in ostriches. Typical antelopes have plantaris muscle fibres only about 5 mm long and camels have no plantaris muscle fibres at all, but only a continuous tendon running from the femur to the toes. In these animals the muscular function of the plantaris has largely or completely disappeared but the elastic function remains. Small mammals such as kangaroo rats (*Diplodomys*, about 100 g) seem to save very little energy by elastic storage: the tendons of their ankle extensor muscles are too thick to make large savings.

Running quadrupeds

Most bipeds hop (like kangaroos) or run (like people). Quadrupeds use a much wider range of running gaits, some of which are shown in Figure 5.9 (Hildebrand, 1976, 1977; Gambaryan, 1974). In the amble, trot and pace, the left and right feet of each pair move half a cycle out of phase. They are described for this reason as symmetrical gaits—the other gaits shown in the figure are asymmetrical. Mammals of different sizes change gaits at different speeds, but generally at about the same Froude number. Bipeds generally change from a walk to a run at a Froude number of about 0.6; quadrupeds generally change from a walk to a symmetrical run at a Froude number of about 0.6 and to an asymmetrical run at some Froude number between 1 and 5.

Many quadrupeds, including horses, trot at low running speeds and use two asymmetrical gaits at successively higher speeds: they canter at moderate speeds and gallop at high speeds. Figure 5.9 shows that in a typical trot, footfalls occur at only two phases of the stride (0 and 0.5). In the canter, footfalls occur at three phases (about 0, 0.3 and 0.7) and in the gallop at four. This is why the sounds made by trotting, cantering and

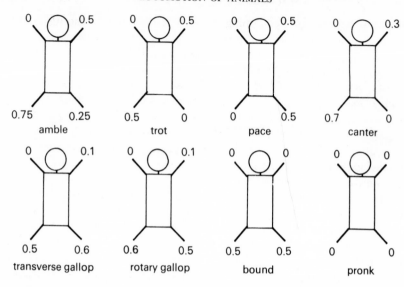

Figure 5.9 Diagrams of quadrupedal running gaits showing typical relative phases of the feet. Many mammals use the trot, canter and one or both of the forms of gallop. Camels pace instead of trotting. Squirrels bound instead of galloping. Some antelopes pronk when alarmed. The amble is the usual running gait of elephants.

galloping horses have quite different rhythms. Some mammals use other running gaits. Camels use the pace instead of the trot, possibly because they have long legs and large feet. It may be easier to avoid putting one foot in the way of another if the fore and hind feet of the same side move in phase. Elephants generally use the amble instead of the trot and have no faster gaits. The relative phases of the amble are the same as for a typical quadrupedal walk.

Equation 5.4 has been used to estimate the energy costs of different quadrupedal gaits (Alexander, Jayes and Ker, 1980). The results seem to show that at low running speeds, the amble is slightly more economical than the trot or pace which in turn are more economical than the gallops. At high speeds nearly all gaits (with the same duty factor) are about equally economical but the bound and pronk are more expensive. The mathematical model fails to explain why most mammals trot at low running speeds and gallop at high speeds. One of its faults is that it treats the animal as a two-dimensional system and so fails to take account of transverse movements and rolling movements. A three-dimensional model might show that the trot is more economical than the amble for slow running. The model does not take account of the bending of the back which occurs in galloping (Figure 5.11f).

Equation 5.4 ignores the masses of the legs. This is justifiable at low speeds but in fast running the positive and negative work done accelerating and decelerating the legs may be as large as the work calculated from the equation. It is the same for all gaits (with given speed and duty factor), so it need not be considered in discussions of the relative merits of different gaits. Little of it is likely to be saved by elastic storage because the muscles that swing the legs forward and back do not have long tendons. Also the legs of dogs and antelopes swing forward, in fast galloping, more slowly than they swing back. If the swinging were maintained elastically, the kinetic energy associated with the movement of the leg relative to the trunk would have to be the same for the forward and backward swings.

To keep the energy needed for swinging the legs as small as possible, the moments of inertia of the legs about the hip or shoulder should be small. Light feet will be advantageous. Ungulates (especially antelopes) have light feet, but other mammals that use their feet for purposes other than running (for instance, for killing prey) necessarily have heavier feet. It is only at high speeds that light feet give much advantage. Measurements have been made of the rates of oxygen consumption of gazelles (with light feet) and cheetahs (with relatively heavy feet) running on a treadmill. The cheetahs were a little heavier than the gazelles but had legs of the same length. At all speeds up to 6 m/s the two species used oxygen at about the same rate, per unit body mass. At higher speeds, however, running depends partly on anaerobic respiration, so meaningful comparisons could not be made.

Most horse races over distances up to 1 mile (1.6 km) are won at 15–17 m/s. Most greyhound races are won at 15–16 m/s. Wild mammals may not be as fast as racehorses and greyhounds which have been selected specifically for speed, possibly to the detriment of other qualities, and many published estimates of the speeds of wild animals are probably exaggerated. More reliable measurements showed that ten species of East African ungulate, chased by a vehicle, ran at maximum speeds between 7 m/s (buffalo) and 14 m/s (zebra and Thomson's gazelle).

Energetics

Rates of oxygen consumption have been measured for many species of mammal, running on treadmills. It has been found in nearly every case

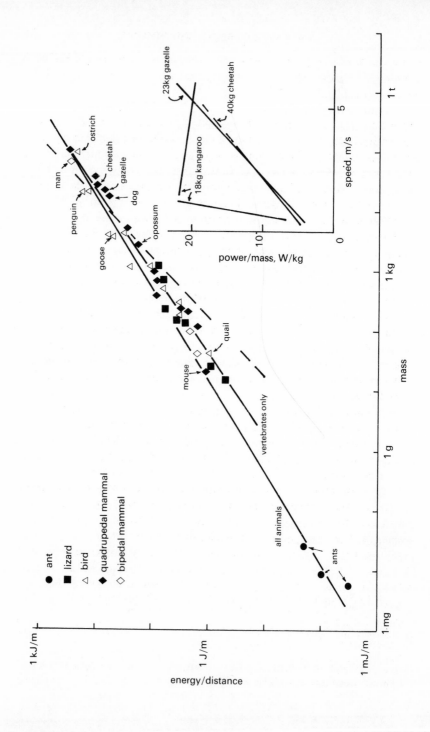

that the metabolic power consumption $P(u)$ is related to the speed u by the equation

$$P(u) = P(0) + au \qquad 5.7$$

where $P(0)$ and a are constants for each individual animal. The equation generally holds throughout the range of walking and running speeds at which oxygen balance can be maintained. This is illustrated by the inset in Figure 5.10. The only important exceptions discovered so far are kangaroos, which use far more power than other animals of similar size at low speeds, at which they shuffle along on all four legs and the tail, but need no more power to hop fast than to hop slowly.

The intercept $P(0)$ in equation 5.7 is generally 1.5 to 2 times the resting metabolic rate, probably because energy is needed for maintaining posture. The gradient a represents the energy needed to travel unit distance. It is larger for large animals than for small ones but does not increase in proportion to body mass—it tends to be about proportional to (body mass)$^{0.7}$. There are also differences between animals of equal mass: for instance penguins, which walk awkwardly, have higher a values than dogs or gazelles of the same mass. There are no consistent differences between mammals, birds and lizards or between quadrupeds and bipeds. Ants have values just a little higher than would be expected by extrapolation from vertebrates.

It might be supposed that a represents the metabolic energy needed to do the mechanical work of running, which can be calculated from force platform records, using equation 5.4, and assuming the efficiencies given on p. 4. Attempts to calculate a in this way give excessively high values for large animals and excessively low ones for small animals. The discrepancy for large animals can be explained by the energy-saving effect of elastic storage: some of the positive and negative work that would otherwise have to be done by muscle is done by tendons and other elastic elements. The discrepancy for small animals has not so far been explained satisfactorily. It has been pointed out that energy is needed to activate

---

Figure 5.10   Graph on logarithmic coordinates of the energy needed to travel unit distance (a, equation 5.7) against body mass, for running animals. The regression line for vertebrates only is from M. A. Fedak and H. J. Seeherman (1979) *Nature* **282**, 713–716. The line for all animals is from T. F. Jensen and I. Holm-Jensen (1980) *J. comp. Physiol.* **B137**, 151–156. The points show data from these papers, and others cited by them. The broken line shows values of a calculated from rates of performance of mechanical work, as explained in the text.

(Inset) Graphs of metabolic power consumption against speed. Data from W. R. Dawson and C. R. Taylor (1973) *Nature* **246**, 313–314 and C. R. Taylor, A. Shkolnik, R. Dmi'el, D. Baharav and A. Borut (1974) *Am. J. Physiol.* **227**, 848–850.

muscles and maintain tension in them even when they do no work, and it has been suggested that this could be the basis of an explanation, but no quantitative explanation has yet been produced (Taylor, 1980).

Stance

Running animals of different kinds stand in different ways, as Figure 5.11 illustrates. Arthropods, newts and reptiles stand with their left and right feet well out on either side of the body. Mammals and birds stand with their feet close together under the body. Small mammals and birds stand and run on relatively bent legs (Figure 5.11e) but large mammals and ostriches stand and run on straighter legs (Figure 5.11g).

Small terrestrial arthropods would be apt to be blown over by the wind if they stood like mammals. This is because they are small and so have large ratios of surface area to weight. Imagine a cylindrical animal with the same density as water. The drag coefficient based on frontal area would be about 1. Using equation A.15, the drag exerted on it by a gentle breeze of 5 m/s, blowing at its side, would be 0.6 times body weight for an insect-sized animal of diameter 3 mm but only 0.006 times body weight for a sheep-sized animal of diameter 300 mm. The insect must stand with its left and right feet far apart (Figure 5.11b), because it will be overturned if the resultant of lift and drag passes above the downwind foot (although this is not necessarily true if the feet adhere to the ground). Nearly all insects stand as shown in Figure 5.11a,b, except the Pauropoda, which stand with their feet closer together, but they live in soil crevices and among leaf litter where wind is not a problem. Most mammals stand with their feet close together as shown in Fig. 5.11e, and can do so because drag is generally small compared to their weight.

Reptiles are too large to be in much danger of being blown over in most circumstances. Their stance with the feet far apart may nevertheless be very convenient for those that climb trees or scramble over rocks, because it makes it relatively easy for them to keep their balance on sloping or irregular surfaces. Large reptiles are less likely to climb and crocodiles often adopt a mammal-like stance for walking.

Reptiles generally stand with the humerus and femur projecting laterally from the body. These bones swing forward and back in roughly horizontal planes when the animals walk (Rewcastle, 1980). Mammals generally move the long bones of their legs in roughly vertical planes, but small mammals such as the opossum (*Didelphis*), rat and ferret have the humerus and femur more nearly horizontal than vertical during most of the stride

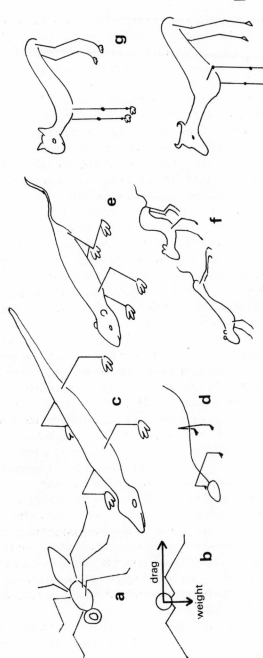

Figure 5.11 Diagrams of animals standing and (d), (f) running. (a), (b), an insect; (c), (d) a newt or lizard; (e), (f) a small mammal and (h) an ungulate.

(Figure 5.11e). This is not always easy to see, but is shown clearly in X-ray cine films (Jenkins, 1971). Cats and dogs keep the humerus and femur more nearly vertical (Figure 5.11g). Large ungulates keep most of the long leg bones quite nearly vertical, but hold the humerus at a large angle to the vertical (Figure 5.11h).

The stances of reptiles and small mammals, with bent legs, have the advantage for climbing that has already been noticed. The stance of small mammals may also assist them to accelerate or jump suddenly, by extending their legs. These stances would, however, be expensive of energy for large animals. Imagine two geometrically similar animals of different size, standing in similar postures. The forces required in leg muscles, to support their weights, would be proportional to body mass. The cross-sectional areas of active muscle would therefore be proportional to body mass, and since the lengths of muscle fibres would be proportional to (body mass)$^{1/3}$, the volumes of active muscle would be proportional to (body mass)$^{4/3}$. If the muscles of the two animals had identical properties the metabolic power needed to maintain tension would be proportional to (body mass)$^{4/3}$. It is more likely that the large animal would have slower muscles, since large animals tend to use lower stride frequencies than small ones. Slow muscles maintain tension more economically than fast ones so the exponent of 4/3 is probably too high: 1.1 or 1.2 would probably be more realistic. Nevertheless, the metabolic power required for standing would be much larger compared to resting metabolism, in the larger animal. The resting metabolic rates of related animals tend to be approximately proportional to (body mass)$^{0.75}$. The moments about joints and so the energy needed for standing are reduced by straightening the legs. It is clear from the structure of their joints that dinosaurs and other large extinct reptiles stood like large mammals, with their long leg bones nearly vertical. It is not clear why ungulates have a near-horizontal humerus, but since it is short it does not imply very large moments at joints.

Mammals of different sizes are not in fact geometrically similar, but the deviations do not affect the above argument much. Measurements on mammals ranging from 3 g shrews to a 3 tonne elephant showed that the lengths and diameters of homologous leg bones tend to be about proportional to (body mass)$^{0.35}$. Ungulates and primates tend to have longer legs than other mammals of equal mass, ungulates because they have very long metapodials (cannon bones) and primates because they have a long humerus, ulna, femur and tibia.

Newts and lizards bend their trunks from side to side when they run (Figure 5.11d); mammals bend their trunks dorso-ventrally when they gallop (Figure 5.11f). In both cases the movements enhance the forward and backward movements of the feet, relative to the centre of mass of the body. They may make higher speeds possible by involving back muscles as well as leg muscles in running, but they may increase the energy cost of running. Mathematical models suggest that less power would be needed for running at given speed if the back were kept rigid, but the role of the back movements is not fully understood.

Climbing

Figure 5.12a shows a mammal climbing a tree trunk. Its weight W is matched by an equal, upward force on its feet, obtained either by friction or by digging in its claws. The weight and the upward force exert a clockwise moment on the body, which cannot therefore be in equilibrium under them alone. Horizontal components of force must act on the feet, as shown in the diagram. The hindfeet push on the tree trunk but the forefeet pull on it. The same principle applies to birds (Figure 5.12c). Woodpeckers rest the tail against the tree trunk as illustrated so that it serves the function of the hindfeet of climbing mammals.

There are several ways in which the forefeet of mammals could exert the necessary pull on tree trunks shown in Figure 5.12a. They are not sufficiently adhesive to grip by adhesion, but squirrels, tree shrews and birds have claws and can dig them into the bark. Most monkeys (except the marmosets, Callithricidae), and some other primates, have no claws. Even without claws, the pull on the tree trunk is easily exerted if the arms are long enough to reach round to the other side. It can also be exerted by friction in some cases, even if the arms cannot reach as much as halfway round. This is shown in Figure 5.12d. Notice that the resultant of the normal force N and the frictional force F acts away from the animal if

$$\theta + 2\phi > 180° \qquad 5.8$$

The frictional force F cannot be greater than μN, where μ is the coefficient of friction. Hence $\phi \leqslant \arctan \mu$. The larger the coefficient of friction, the smaller the angle θ required for climbing.

The hands and feet of primates have remarkably high coefficients of friction with smooth surfaces (Cartmill, 1979). Figure 5.12d shows an experiment in which animals were put on a smooth plywood board which was tilted until they slid off. Sliding occurred when the angle ϕ reached $\arctan \mu$. People supporting themselves on the palms of their hands and the soles of their feet started sliding before ϕ reached $45°$ ($\mu < 1$) but

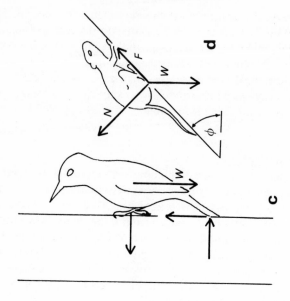

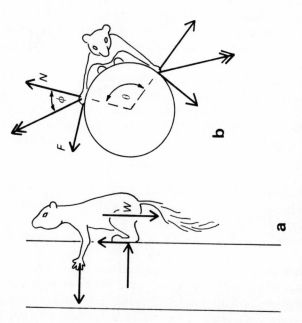

Figure 5.12 (a), (b) Diagrams of a mammal climbing a tree trunk.
(c) A bird climbing a tree trunk.
(d) A mammal resting on a smooth slope.

smaller primates did better. Infant mouse lemurs (*Microcebus*) did not slide until ϕ reached 72.5° ($\mu = 3.2$). In contrast, an infant squirrel (*Sciurus*) slipped at 35°, though it was capable of climbing vertically up the experimenter's clothes. Its claws could not obtain a purchase on the smooth wood. (Infants were used for these experiments because adults jumped off before they started slipping.)

Squirrels, tree shrews and many monkeys often run along the tops of branches, on all fours. This is not easy to do if the diameter of the branch is small compared to the diameter of the body, but monkeys have prehensile hands and feet, and some have tails that can be used for grasping small branches. Birds and chameleons also have prehensile feet. Some monkeys and apes spend a lot of time climbing among small branches, grasping several branches with different limbs and using branches above as well as below them. A few of them, such as gibbons (*Hylobates*) brachiate, travelling by swinging by the arms alone from overhead branches.

Flies and geckoes can run upside-down on the ceilings of rooms. *Gekko*

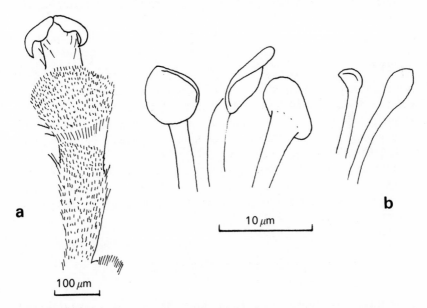

Figure 5.13 The foot of a ladybird beetle, *Anisosticta*. (a) shows a complete foot and (b) shows a few of the setae at higher magnification. (Based on scanning electron micrographs by N. E. Stork (1980) *Zool. J. Linn. Soc.* **68**, 173–306.)

and *Anolis* (another lizard) can climb vertical sheets of new, very carefully cleaned plate glass, but fall off if the glass is tilted beyond the vertical. They can move about more freely on slightly dirty glass, and even hang upside-down from it. *Rhodnius* (a bug) can remain attached to polished plate glass when it is vertical or even tilted a little beyond the vertical (but slips on Teflon plastic at only 40° to the horizontal).

The feet of these animals must have adhesive properties. The mechanism of adhesion has not been fully explained, but seems to be the same for lizards and insects (Stork, 1980). *Gekko, Anolis, Rhodnius* and many beetles that run on the smooth surfaces of leaves have very similar adhesive pads on their feet. They are covered by a dense pile of setae, fine hair-like projections with broader flattened ends (Figure 5.13). In *Gekko* these ends are only 0.2 μm wide (too narrow to be resolved by light microscopy and so they have been studied by electron microscopy) but in insects they are considerably larger.

The fine, flexible setae can make very close contact indeed with any solid surface, whether smooth or rough. It has been suggested quite plausibly that the mechanism of adhesion may depend on the intermolecular forces known as van der Waals' forces, which are explained in textbooks on chemistry.

Jumping

A wide variety of animals make impressive jumps: bushbabies (*Galago*) jump from one branch to another, fleas reach the legs of their hosts by jumping, locusts jump as a preliminary to flight and flea beetles seem to use jumping as a means of escape from danger. In general, mammals jump higher than insects (Figure 5.14). Should this be expected?

Suppose an animal of mass m takes off vertically from the ground with velocity v_0 and jumps to a height h. As it leaves the ground it has kinetic energy $\frac{1}{2}mv_0^2$. At the top of its jump it has no kinetic energy but has gained potential energy mgh. If all the kinetic energy has been converted to potential energy,

$$\frac{1}{2}mv_0^2 = mgh$$
$$h = v_0^2/2g$$

5.9

Compare two animals of different sizes with the same proportion of muscle in their bodies. The work the muscles can do in a single contraction to give the body kinetic energy and start the jump, is likely to be

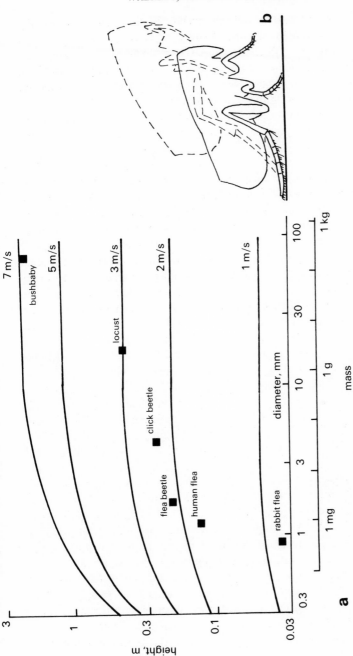

Figure 5.14 (a) Graphs calculated from equation 5.10, showing the heights to which spheres of different sizes would rise if projected vertically with the initial velocities shown. The spheres are assumed to have the density of water. The points show heights to which animals of various masses can jump. (Data from H. C. Bennet-Clark, 1977 in T. J. Pedley (edit.) *Scale Effects in Animal Locomotion* 185–201, London, Academic Press, and sources therein.)
(b) A flea (*Spilopsyllus*) taking off for a jump (based on drawings from films by H. C. Bennet-Clark and E. C. A. Lucey, 1967, *J. exp. Biol.* **47**, 59–76).

proportional to m. Therefore animals of different sizes should be able to accelerate to the same take-off velocity and jump to the same height.

That argument was too simple because it ignored air resistance (Bennet-Clark and Alder, 1979). The initial kinetic energy is not all converted to potential energy but some is used doing work against drag. At time t during its jump, let the animal have velocity v and suffer drag D. Its acceleration using equation A.15 is given by

$$dv/dt = -g - (D/m)$$
$$= -g - (\rho S_f C_D v^2/2m) \qquad 5.10$$

(Here ρ is the density of the air, S_f is the frontal area of the animal and C_D is the drag coefficient.) The equation can be solved to discover the height of the jump, given the take-off velocity. This has been done for a range of imaginary spherical animals to obtain Figure 5.14a. This graph shows that if a large animal and a similar small one take off with the same vertical velocity, the small one does not rise so high. The reason is that S/m is larger for the small animal: S is proportional to (diameter)^2 and m to (diameter)^3.

Vertebrate skeletal muscle can do work up to a maximum of about 200 J/kg in a single contraction. Bushbabies (*Galago senegalensis*) have about 36 % muscle in their bodies and might be expected to be able to do 72 J/kg body mass when they jump. However, the muscles cannot do so much work if they contract fast or through less than the maximum possible range of lengths. Also, not all of the body muscles can be used for jumping. Bushbabies can jump to a maximum of about 2.26 m which implies a take-off velocity of 6.7 m/s and a kinetic energy of 22 J/kg. Few other animals can jump so high. (A human high jumper starts with his centre of mass about 1 m from the ground and has to raise it little more than 1 m to clear 2 m, so his performance is much less impressive.)

A flea with enough muscle in its body could presumably take off as fast as a bushbaby, with a velocity of nearly 7 m/s. Figure 5.14a shows that it would probably not rise higher than about 0.5 m, less than one quarter the height of the bushbaby's jump. Jumping insects have relatively smaller proportions of muscle available for jumping, and take off at much lower velocities than bushbabies. They rise less high, but waste less energy against drag than if they took off faster. A small flea (*Spilopsyllus*, the rabbit flea) takes off at only 1 m/s and jumps a height of 35 mm. Human fleas (*Pulex*) take off faster and rise higher, to at least 130 mm (Bennet-Clark and Lucey, 1967).

Most animals jump by extending their legs suddenly (Figure 5.14b). An animal accelerates itself from rest to its take-off velocity v_0 over a distance l which cannot exceed the length of its legs. From equation A.4

$$l \geqslant \tfrac{1}{2}v_0 t$$
$$t \leqslant 2l/v_0$$

5.11

where t is the time taken for extension of the legs. For a bushbaby, l is about 0.1 m, so if it takes off at 7 m/s t is about 30 ms. For a rabbit flea l is about 0.4 mm, so take-off even at only 1 m/s gives t the very small value of only 0.8 ms.

No known muscle can make a single complete contraction in a time as short as this. Mosquitoes beat their wings at about 600 Hz, so each muscle contraction lasts about 0.8 ms, but these are fibrillar muscles in a state of oscillation. Isolated contractions cannot be made so fast.

The jumps of fleas and other small insects are made possible by catapult mechanisms. Elastic strain energy stored in a catapult by a slow muscular contraction can be released very rapidly indeed in an elastic recoil. In fleas, the principal elastic elements of the catapult are blocks of the protein resilin, and may have evolved from blocks of resilin in the wing mechanism of flea ancestors. (The fleas themselves have no wings.) In locusts, the principal elastic elements are the apodeme (tendon) of the "knee" extensor muscle and part of the knee joint (Bennet-Clark, 1975). Fleas, locusts and other jumping insects have trigger mechanisms which enable them to fire their catapults suddenly.

CHAPTER SIX

CRAWLING

The preceding chapter was about walking and running. Animals may use other methods for moving over solid surfaces and for burrowing through solid media. Several basic mechanisms—two-anchor crawling, peristalsis, pedal waves, serpentine and amoeboid crawling—are considered in turn in this chapter. Some of these are further discussed in relation to soft-bodied animals by Trueman (1975).

Two-anchor crawling

The caterpillar shown in Figure 6.1a has true legs at its anterior end and prolegs at its posterior end. With the prolegs fixed it extends its body, moving its anterior end as far forward as possible. Then with the true legs stationary on the ground it bends its body to bring the prolegs as close to them as possible. The anterior end of the body is moved forward while the posterior end is anchored, and vice versa. In this case the legs and prolegs are the anterior and posterior anchors. Leeches (Figure 6.1b) have suckers which serve as anchors, but in other respects the motion of the leech is like that of the caterpillar.

Figure 6.1c shows a different application of the two-anchor principle, for burrowing (Trueman, 1967). Bivalve molluscs have the two valves of the shell connected by an elastic hinge. Contraction of the adductor muscle closes the shell, but when this muscle relaxes, the elastic hinge makes the shell spring open (Figure 1.2a). Typical bivalves are filter feeders that live buried in sand or mud. If they are dug up, they can bury themselves again. Some (such as *Ensis*) burrow deeper when the surface of the sand or mud is exposed at low tide, and climb near the surface again as the tide rises. A West Indian species (*Donax denticulatus*) lives in the surf, emerging from

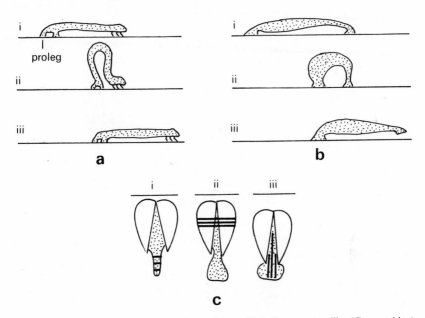

Figure 6.1 Diagrams showing successive positions of (a) a looper caterpillar (Geometridae) crawling; (b) a leech crawling and (c) a bivalve mollusc burrowing. Thick lines in (c) represent muscles that are active at each stage.

the sand from time to time and allowing the waves to move it up or down the shore as the tide rises or falls. It can bury itself very fast, in only 4 s, but most bivalves burrow much more slowly (for example, a British species of *Donax* takes about 60 s to bury itself.)

The movements shown in Figure 6.1c are best seen by allowing a bivalve to burrow in a narrow glass container filled with wet sand or mud. In stage (i) the adductor muscle is relaxed and the shell has partially opened, jamming itself tightly against the surrounding sand. The muscular foot is probing the sand, pushing down as deep as possible. This probing is effected by muscles running across the foot which tend to make it longer and thinner. In stage (ii) the adductor muscle has contracted, closing the shell and driving blood out of the main part of its body into the foot. The shell is no longer anchored because it has closed but the foot has been inflated by the influx of blood and is now anchored. Closing of the shell also drives water out of the mantle cavity, tending to squirt away the sand from immediately under the shell. In stage (iii) the foot has shortened, pulling the shell down into the cleared space.

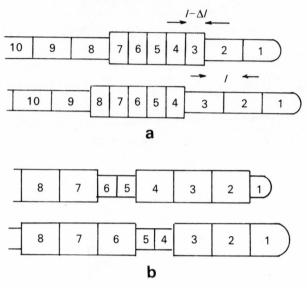

Figure 6.2 Diagrams of successive stages of crawling by (a) an earthworm such as *Lumbricus*, and (b) a worm such as *Polyphysia*.

Peristalsis

Figure 6.2a shows diagrammatically how earthworms crawl. The body is segmented with internal partitions (septa) crossing the body cavity between segments, so that each segment is a separate, fluid-filled unit. Each segment can be made long and thin by contraction of its circular muscles or short and fat by contraction of longitudinal muscles, but its volume remains constant, apart possibly from slight leakage of fluid through the holes in the septa for the nerve cord. (There are muscles to keep these holes closed.)

The figure shows waves of lengthening and shortening travelling backwards along the worm. Each segment is shown (for simplicity) fully lengthened or fully shortened, but it would have been more realistic to have shown gradations between the extremes. While the segments are short and thick, they are anchored. Because they are thick, most of the worm's weight rests on them so that they are anchored by friction. In addition, bristle-like chaetae are protruded from the thickened segments. Between stages (i) and (ii) the wave has moved one segment posteriorly. Segment 3 has lengthened, pushing the more anterior segments forward. Segment 8 has shortened, pulling more posterior segments forward.

The speed of crawling depends on several factors. Let each segment have length l while elongated and $(l - \Delta l)$ while shortened, and let a fraction q of the segments be elongated at any instant. Let the waves of contraction travel along the worm with speed U. Between stages (i) and (ii), the wave has moved a distance l along the worm, in time l/U. Segment 1 has moved forward Δl (due to elongation of segment 3) so its speed is $U \cdot \Delta l/l$. It has this speed for a fraction q of the time, but is shortened and stationary for the rest of the time, so the mean speed of this segment (and of the whole worm) is $q U \cdot \Delta l/l$. The speed of crawling can be increased by making the waves travel faster (increasing U), by making the segments lengthen and shorten more (increasing $\Delta l/l$) or by increasing the fraction q of the segments that are extended at any instant. Since $\Delta l/l$ and q must both be less than 1, the forward speed of the worm relative to the ground must be less than the backward speed U of the waves relative to the worm. The waves must travel backwards relative to the ground as well as relative to the worm. Earthworms crawl at speeds of about 5 mm/s.

This method of crawling can also be used by animals that are not divided into closed segments, provided they can make parts of the body long and thin while others are short and fat, and so it is used by unsegmented nemertean worms as well as by segmented earthworms. It is suitable for burrowing as well as for crawling, as the shortened, thickened segments are well anchored by being jammed against the surrounding soil.

Figure 6.2b shows an alternative method for crawling by peristalsis (Elder, 1973). It is used by some polychaete worms that are segmented but have no septa (except between a few segments at the ends of their bodies), so that fluid can move freely from one segment to another. It is also used by some holothurians (sea cucumbers), which are not segmented. The circular and longitudinal muscles of each segment (or part of the body) contract simultaneously, driving fluid out of the segment, and relax simultaneously, allowing fluid to be driven in from other segments. Between the two stages shown in Figure 6.2b, segment 4 has contracted, pulling segment 5 forward, and segment 6 has expanded, pushing segment 5 forward. Notice that in this technique of crawling, the waves of contraction move forward along the animal. This is called *direct peristalsis*; the technique used by earthworms (in which the waves travel backwards) is called *retrograde peristalsis*. In both cases the segments are anchored while their diameter is greatest.

Direct peristaltic crawling seems to be used only by animals that live in or on very soft submerged mud. It has been suggested that this technique is particularly suitable for soft mud because it enables the worm to move reasonably quickly while keeping a large area of the body wall anchored. The mathematical analysis above applies to direct as well as to retrograde peristalsis. Note that q is the fraction of thin, unanchored segments. In retrograde peristalsis the anchored segments are shortened but in direct peristalsis they are elongated, so for any given value of q, an animal using direct peristalsis has its anchorage distributed over a larger area. A smaller value of q would increase the area in either type of crawling but would also reduce crawling speed for any given wave speed.

The anterior end of an animal burrowing by either method of peristalsis moves forward only while the anterior segments are slender. Earthworms probably slide the slender anterior segments into existing crevices in the soil. The crevices are enlarged by subsequent swelling of the segments, when their longitudinal muscles contract. If large forces are required, high pressures must develop in the body cavity. A segmented animal can confine the high pressure to a few anterior segments, but if worms without septa extended their burrows in the same way the high pressure would have to be developed throughout the body cavity and much more muscle would have to be activated. The worm *Polyphysia*, which burrows by direct peristalsis, does not use pressure to enlarge crevices. Instead it extends its burrow by making scraping movements with its anterior end.

It is difficult to measure pressures in animals while they are burrowing— no-one has succeeded so far in connecting a pressure transducer to a burrowing animal without restricting its freedom to burrow. Such measurements as have been made (Seymour, 1969) tend to confirm the account given above. Pressures up to 2.5 kPa have been recorded from the body cavities of earthworms burrowing in loose earth, but much higher pressures are probably developed in firmer soil. Pressures up to 7.5 kPa have been recorded from worms squirming in air and it has been estimated that pressures up to 40 kPa may be possible in burrowing. (In air the pressures are limited by the capability of the relatively weak circular muscles, but the longitudinal muscles need no assistance from the circular ones in enlarging a crevice). Pressures up to 10 kPa have been recorded from the haemocoel of the bivalve *Ensis* while it was burrowing in sand by the technique shown in Figure 6.1c.

Pedal waves

Snails, other gastropod molluscs, and chitons, crawl on a large "foot". If they are watched from below while crawling on glass, waves of muscular activity can be seen travelling along the sole of the foot. In chitons the waves are retrograde (travel backwards) but in the snail *Helix* they are direct (travel forwards). In these cases the muscular activity is in phase across the whole width of the foot but in other cases waves travelling along the left side of the foot are half a cycle out of phase with waves on

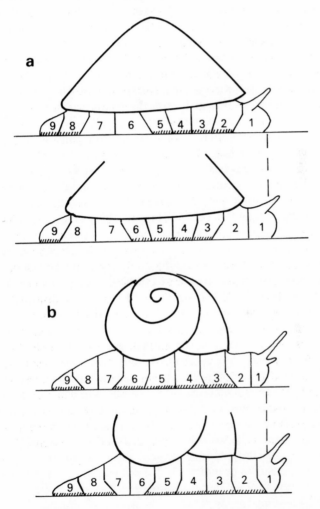

Figure 6.3 Diagrams of two successive stages of crawling by (a) a mollusc that uses retrograde pedal waves, such as a limpet (*Patella*), and (b) a mollusc that uses direct pedal waves such as a snail (*Helix*). The anchored parts of the foot are indicated by hatching, in each case.

the right. This is true of winkles (*Littorina*) and limpets (*Patella*), which have retrograde waves, and also of top shells (*Gibbula*), which have direct waves.

If marks are made on the sole of the foot before the animal is put on glass, the movements involved in crawling are more easily seen. Each mark moves forward at one stage in the passage of a wave of muscular activity, and remains stationary at another. It can be concluded that movement is effected by waves of lengthening and shortening of the foot, like the waves involved in peristaltic crawling of worms. The waves, which affect the whole body in worms, are here confined to the foot (Figure 6.3). If the shortened parts of the foot are anchored, retrograde waves are needed to move the snail forwards (Figure 6.3a). If the lengthened parts are anchored, direct waves are required (Figure 6.3b).

The feet of molluscs secrete slimy mucus, which consists of water and dissolved salts with a small proportion of glycoprotein (3–4 % in the slug *Agriolimax*). The mucus acts as glue, attaching the foot firmly to any solid surface. This makes aquatic snails less likely to be dislodged by water movements and enables land snails to climb up the plants they feed on.

It used to be thought that crawling involved lifting the foot, part by part, and setting it down; the lifted parts were free to move forward but the parts resting on the substrate were anchored. This hypothesis is implausible because the whole foot lies very closely on the substrate. Even if there were no glue-like mucus, large forces would be needed to lift the foot from the surface of (for instance) a rock, for the same reason that makes it difficult to separate two damp microscope slides. The mucus makes the attachment even firmer. It is well known that it is hard to pull limpets off rocks.

A different mechanism that may attach parts of the foot and release others has been suggested by Denny (1980). It is based on experiments with mucus like the one shown diagrammatically in Figure 6.4a, although the apparatus was not quite as simple as the diagram suggests. In the diagram, a layer of mucus is sandwiched between two flat plates. The lower plate is held stationary while the upper one is moved to the right at constant speed, and the force F required to move it is measured. If the mucus were an elastic solid that behaved according to Hooke's Law, the force would increase in proportion to the distance moved by the plate, and so in proportion to time, but if the mucus were a viscous liquid the force would be constant for as long as the speed of movement remained constant (Figure 6.4b). The results of actual experiments were like Figure 6.4c. The mucus behaved like an elastic solid until the force reached a certain value, at which the mucus yielded. When it yielded the force fell to a value that remained constant: the mucus behaved like an elastic liquid. When movement stopped, the mucus quickly (within a second) reverted to its original solid properties.

The passage of muscular waves along the foot exerts forward forces on

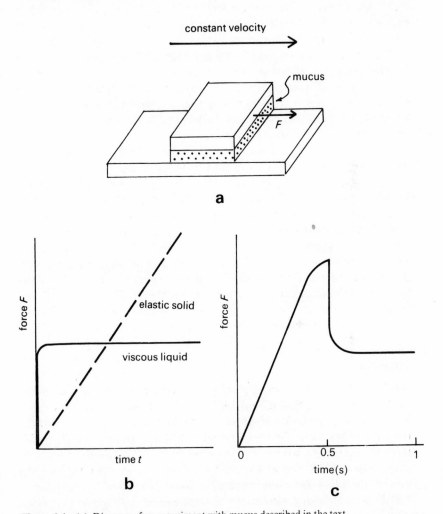

Figure 6.4 (a) Diagram of an experiment with mucus described in the text.

(b) Graphs of force F against time t such as would be obtained in the experiment, if the mucus was an ideal elastic solid or viscous liquid.

(c) A graph showing the result of an experiment with slug mucus by M. Denny (1980) *Nature* **285**, 160–161.

some parts of the mucus and backward forces on others. The snail must be nearly in equilibrium (otherwise it would have a large acceleration) so the forward and backward forces must be about equal. They may however be applied to different areas of mucus, so that the forward and backward shear stresses are different (shear stresses are explained in the appendix). In Figure 6.3a the lengthened parts of the foot occupy a smaller area than the shortened parts, so greater shearing stresses are developed under them. If the forces are large enough the mucus under the lengthened parts yields but the mucus under the shortened parts does not. The shortened parts remain anchored, and the retrograde waves move the limpet forwards. In Figure 6.3b, however, the shortened regions occupy less area than the lengthened ones. Yielding occurs under the shortened regions and the direct waves move the snail forwards.

Pond snails such as *Lymnaea* crawl by means of cilia on the foot, rather than by muscular waves. Triclad flatworms crawl slowly by means of cilia but use muscular waves as well to crawl faster. All these styles of crawling are slow. Snails (*Helix*) and limpets (*Patella*) crawl at about 2.5 and 1 mm/s, respectively. *Lymnaea peregra* crawls at up to about 3 mm/s.

A form of pedal locomotion is used by snakes such as boas, which can crawl along with the body straight (Gans, 1962). The skin of the ventral surface is moved relative to the rest of the body by muscles that pull parts of it backward and parts forward, so that retrograde waves travel along the body. Anchorage depends on friction, not mucus. The lengthened parts of the skin are lifted clear of the ground so that the snake's weight rests on the shortened parts which are thereby anchored.

Serpentine crawling

Snakes more usually crawl by passing waves of bending along their bodies as shown in Figure 6.5a (Gans, 1962). They form their bodies into bends around stones, tussocks of grass or anything else that can provide a purchase. They make the bends travel backwards along the body and so push themselves forwards. The principle is the same as for undulatory swimming as illustrated in Figure 2.2: movements at right angles to the axis of the body are more strongly resisted than axial movements. In undulatory swimming, movements at right angles to the axis occur, as well as axial movements, and the waves of bending travel backwards relative to the water. In serpentine crawling, movements at right angles to the axis may be prevented by immovable solid objects, and the waves remain stationary relative to the ground.

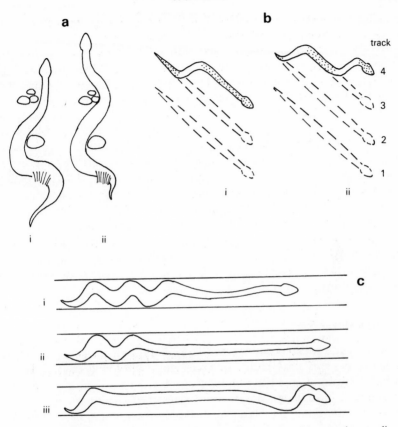

Figure 6.5   Diagrams of three methods of crawling used by snakes: (a) serpentine crawling; (b) sidewinding and (c) concertina crawling.

This technique of crawling is also used by small nematode worms, crawling between particles of soil. These worms, however, use dorsal and ventral instead of transverse bends for crawling.

Rattlesnakes (*Crotalus*) and some other snakes crawl over loose sand by sidewinding, the technique shown in Figure 6.5b. This involves bends travelling backwards along the body but works quite differently from serpentine crawling. In serpentine crawling the body slides along a continuous track. In sidewinding each part of the body is stationary while on the ground but is lifted periodically to a new position. In Figure 6.5b, only the stippled parts of the body are on the ground. The mid-part of the body is being lifted across from track 2 to track 3 and the head is being lifted to a new position to start a new track 4.

Sidewinding depends on the same mechanical principle as peristaltic crawling. The length of a segment of the body has a positive component in the direction of crawling while off the ground, and a negative component while on the ground. In this rather obscure sense it is shorter while on the ground, so a retrograde wave moves the snake forwards.

A method of crawling that is more obviously related to peristalsis is shown in Figure 6.5c. The diagram is based on observations of snakes crawling in glass tubes but the same technique is also used by snakes to climb large trees by ascending grooves in their trunks. The folded part of the body is wedged in the tube. The anterior folds are opening, pushing the snake's head forwards. New folds are being added at the posterior end of the group, pulling the tail forwards. Waves of folding are thus travelling backwards along the body. The length of a segment of the body (measured along the axis of the tube) is less while it is folded and anchored, than while it is extended and moving forward. The retrograde wave therefore moves the snake forwards.

Amoeboid crawling

The well-known amoebas are the large ones such as *Amoeba proteus* (length about 0.5 mm). Their shape is irregular and changes as they crawl, which they do at speeds up to about 5 µm/s. Their cytoplasm has many granules in it, so its movements can be observed under a microscope. In crawling, the outer cytoplasm (ectoplasm) remains stationary while an inner core (endoplasm) flows forward. Since the ectoplasm does not get left behind, this implies that ectoplasm is converted to endoplasm at the rear end of the amoeba and endoplasm to ectoplasm at the front end. The rear end puckers as the endoplasm moves forward out of it, but the advancing front end is rounded.

There has been a great deal of controversy as to how these movements are caused (Allen, 1973). Part of the evidence is the pattern of flow, which is indicated by arrows in Figure 6.6a. Near the front of the amoeba, all the endoplasm flows forward at the same speed. At the rear there are gradients of speed with the fastest flow in the centre, just as fluid flows fastest along the axis of a pipe and slower near the walls. Other evidence has been obtained by polarizing microscopy, which shows that the endoplasm is positively birefringent, implying that its molecules tend to be aligned in the direction of crawling. Also, interference microscopy shows that endoplasm

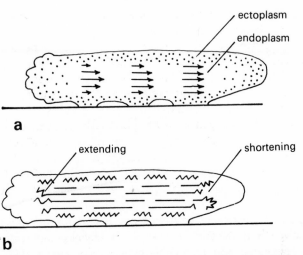

Figure 6.6 Diagrammatic vertical sections of an amoeba crawling: (a) shows the ectoplasm, the endoplasm and (by the lengths of the arrows) the relative speeds of different parts of the endoplasm; (b) illustrates the role of extension and shortening of different parts of the cytoplasm.

contains more water than ectoplasm. It is generally agreed that the explanation of crawling which fits these observations best, is that the endoplasm shortens as it changes to ectoplasm at the front end of the amoeba, pulling the rest of the endoplasm forwards. This is represented in Figure 6.6b as if straight molecules in the endoplasm were shortening to become zigzag molecules in the ectoplasm. It seems more likely that some kind of sliding filament mechanism operates, since thick and thin filaments like those of muscle are present in the cytoplasm (see p. 11). The arrangement of the filaments during crawling is unknown because it seems to be altered by preparation for electron microscopy, which destroys the birefringence of the cytoplasm.

Other amoebas crawl differently. For instance, *Difflugia* crawls by a two-anchor mechanism. This amoeba occupies a heavy shell made of sand grains cemented together. With the shell resting on the substrate, it extends a lobe of its body (pseudopodium) forwards. It attaches the tip of the pseudopodium to the substrate and shortens it, dragging the shell forwards.

CHAPTER SEVEN

LOCOMOTION IN GENERAL

Chapters 2 to 6 considered different types of locomotion separately. This one considers them together. Its first section makes comparisons between running, swimming and flight and between animals of different sizes. Subsequent sections are about prey pursued by predators and about possible advantages of moving around in groups.

Comparisons

The animals compared in this section belong to two notably successful groups, the vertebrates and the insects. These are the only groups to have evolved flight as well as running and swimming. In general, they travel faster than most members of other groups.

Figure 7.1 compares speeds of animals. These are speeds that can be maintained for many minutes without the animal incurring an oxygen debt. It is possible that some of the animals can travel a little faster and still incur no debt, as the observations from which the data come were not designed specifically to discover maximum speeds. It is certain that many of the animals can travel faster for short distances. For instance, the maximum speed that dogs have been shown to maintain aerobically is only 3 m/s but domestic pet dogs of the larger breeds can sprint at around 10 m/s, and greyhounds race at 15–16 m/s.

The figure shows that as a general rule, large animals fly, run and swim faster than smaller ones. Also, when animals of similar size are compared, flight is faster than running which in turn tends to be faster than swimming. Insects fly at about the speeds that might be predicted for animals of their range of sizes, by extrapolation from birds and bats. The running speed shown for ants is lower than would have been predicted by

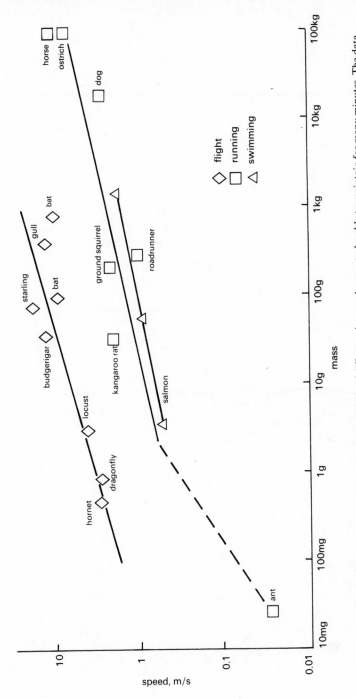

Figure 7.1 A graph showing the highest speeds which animals of different sizes are known to be able to maintain for many minutes. The data come from papers cited in the captions of Figures 4.8 and 5.10, and from J. R. Brett and N. R. Glass (1973) *J. Fish. Res. Bd. Can.* **30**, 379–387 and T. Weis-Fogh (1967) *J. exp. Biol.* **47**, 561–587.

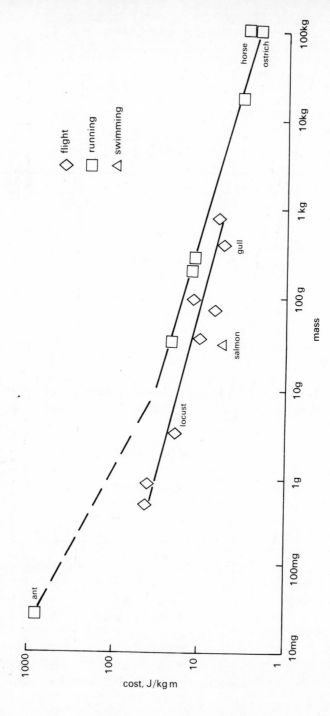

Figure 7.2 A graph showing net costs of transport for animals of different sizes, from the same sources as Figure 7.1.

extrapolation from vertebrates and may well not be a maximum. It is the speed at which ants ran spontaneously in an experiment, and is included only for lack of other data on sustained running speeds of insects.

Energy requirements for locomotion are shown in Figure 7.2 (see also Schmidt-Nielsen, 1972). The net cost of transport is the metabolic energy needed for locomotion (in excess of resting requirements) per unit mass of animal and distance travelled. Thus if an animal of mass m uses metabolic power $P(u)$ when travelling at speed u, and $P(0)$ when resting, the net cost of transport is $[P(u) - P(0)]/mu$. The data have been calculated from measurements of oxygen consumption or (for some of the insects) heat production.

For running mammals, $[P(u) - P(0)]$ is generally about proportional to u (Figure 5.10, inset) so net cost of transport is about the same for all speeds. For flying and swimming animals, cost of transport varies with speed. The data for flying animals refer to the most economical speeds that have been investigated. For fish, net cost of transport is lowest at very low speeds (Figure 2.7b) but the costs of transport given for fish refer to the maximum speed that can be maintained aerobically.

Figure 7.2 shows that in general, larger animals have lower costs of transport than smaller ones. For animals of similar size, flight is more economical than running, and swimming is more economical still. The power $P(u)$ required for flight is very high, but the speed u is also high so the cost of transport is quite low. The costs shown for flight refer to flapping flight in still air, but in favourable conditions some birds travel considerably more economically by soaring.

It is difficult to find information about the distances animals travel in ordinary daily activities, although some animals wander mainly within well-defined areas. Some occupy a territory and defend it against intruders, and this territory may be held by an individual, a pair or a larger group. Other animals confine their movements largely to a home range without claiming exclusive rights there. Figure 7.3 shows that large mammals generally have larger home ranges than small ones (they generally need to search a large area to find sufficient food). Also carnivorous and insectivorous mammals, such as wolves and shrews, generally have larger home ranges than herbivorous ones, such as deer and voles, probably because animal food is more sparsely distributed than vegetable food. Figure 7.3 also shows that the feeding territories of hawks are larger than those of an herbivorous bird of similar size, the ptarmigan (*Lagopus mutus*). They are also larger than the territories of smaller carnivorous and insectivorous birds such as shrikes and warblers. However, these data refer

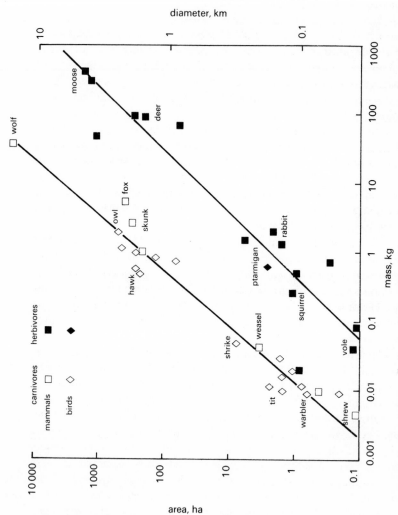

Figure 7.3 A graph showing the areas of the home ranges of mammals, and of the feeding territories of birds. Areas are plotted against body mass. The scale on the right shows the diameters of circles of the same areas as the home ranges or territories. Data are from A. S. Harestad and F. L. Bunnell (1970) *Ecology* **60**, 389–402 and from T. W. Schoener (1968) *Ecology* **49**, 123–141.

to territories and the mammal ones to home ranges, so it would be wrong to attach too much significance to the apparent similarities of area between the ptarmigan and herbivorous mammals, or between carnivorous birds and carnivorous mammals.

Figure 7.3 would have presented a less coherent picture, if it had been extended to show the home ranges of insects. Most of the available data refer to social insects (Urbani, 1979). Not surprisingly, the home ranges of their colonies are larger than would be predicted by extrapolation from mammals, which live in much smaller groups. For instance, mound-building termites such as *Cubitermes* have individual body masses of only a few milligrams but have colony territories of at least $10 \, \text{m}^2$. Honey bees, with masses of about 0.1 g, forage up to several kilometres from the hive.

Many animals make seasonal migrations over distances far longer than their ordinary daily journeys (Baker, 1978). Many birds migrate between the northern and southern hemispheres, spending the summer in each; many warblers migrate between Northern Europe and Southern Africa, and the bobolink (*Dolichonyx*, a passerine bird) migrates between Canada and Argentina. Journeys of over 5000 km each way are made by many species. The northern fur seal (*Callorhinus ursinus*) breeds in the North Pacific in winter but spends the summer up to 5000 km further south. Caribou (*Rangifer*) travel up to 1000 km each way between tundra in summer and forest in winter. Wildebeest (*Connochaetes*) make an annual circular tour of about 1000 km on the Serengeti plains, taking advantage of differences in the dates of the rainy season and the consequent growth of grass.

These journeys are made annually, but some other animals make long journeys just once in each direction. Pacific salmon (*Onchorhynchus*) and sea lampreys (*Petromyzon*) hatch from eggs in rivers but spend much of their lives in the sea and finally return to rivers to breed. Many of the rivers are short, but some salmon breed in the Columbia River more than 2000 km from its mouth. Monarch butterflies (*Danaus plexippus*) breed in the U.S.A. and Canada. Members of the last generation of the summer travel up to 3000 km south, to spend the winter in Florida, Mexico etc. Several generations may occur during the northward return flight in spring.

The longest migrations made by walking animals are much shorter than many migrations made by swimmers and fliers. Walking and running are less economical than swimming and flight (Figure 7.2), so very long migrations are less likely to be advantageous for walkers.

Some migrations involve travelling long distances without food or water or both. Neither is available to birds crossing the Gulf of Mexico (800–2000 km, depending on route), and neither is easily available to birds crossing the Sahara (1400 km). Salmon do not feed on their journey upstream, though they may need a lot of energy to swim against the current.

To gain an impression of what is possible, consider an animal of mass m starting a journey with a store of fat or other food incorporating energy Em. Let it travel at speed u using power $P(u)$. The energy store will last for time $Em/P(u)$, enabling the animal to travel without food for a distance $Emu/P(u)$. This is the energy store per unit body mass, E, divided by the gross cost of transport $P(u)/mu$. Suppose for example that the energy store consists of fat amounting to 25% of the animals' body mass. (Some birds accumulate more fat than this before migrating, but salmon start their upstream migration with only 10–15% fat.) The heat of combustion of fat is 40 MJ/kg so in our example $E = 10$ MJ/kg. If the gross cost of transport were 10 J/kg m the animal could travel 1000 km without food, and if the gross cost were only 5 J/kg m it could travel 2000 km. Figure 7.2 indicates that gross costs of transport in this range are likely for many animals. The figure shows net costs of transport, $[P(u)-P(0)]/mu$, which are smaller than gross costs but are unlikely to be less than half the gross cost for journeys at reasonable speed. For flying birds, $P(0) \ll P(u)$, and the gross and net costs are almost equal.

Escape and capture

Predators chase prey on land, in water and in the air. For instance, cheetahs chase antelopes, pike chase minnows and hawks chase other birds. If the prey can travel faster than the predator it can escape, unless it is taken by surprise. If it is slower than the predator it may be captured, but it may be able to escape by dodging and swerving (Howland, 1974).

Imagine a predator chasing a prey animal along the straight line ABC in Figure 7.4a. The prey is slower than the predator and will be caught if it continues in a straight line. Suppose however that it swerves to the left when it reaches C, and continues at speed u along a circular arc of radius r. When the prey is at C the predator is at B. It has very quick reactions, and starts swerving at the same instant as the prey, travelling at speed U along an arc of radius R. The paths of the predator and prey intersect at D. This figure illustrates a chase with $u = 0.75U, r = 0.5R$: the prey is only three-quarters as fast as the predator but can turn with half the radius. One unit of time after they start swerving the two animals are at the points marked 1, after two units of time they are at the points 2, and so on. The prey animal arrives at D after 6 units of time. The predator would arrive there after only 5.4 units if it continued at constant speed, but by slowing down a little it can intercept and capture the prey at D.

In this case the distance BC is quite large: the prey swerves while the predator is still quite a long way behind it. Figure 7.4b represents another

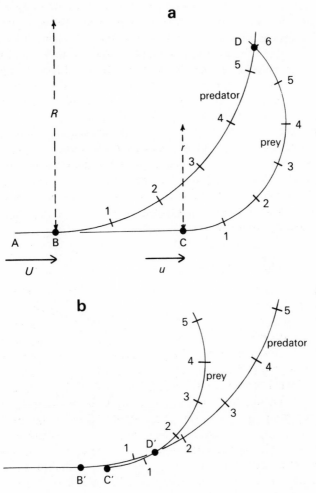

Figure 7.4 Diagrams of a predator chasing a prey animal—further explanation is given in the text.

chase in which the prey delays its swerve until the predator is much closer (when the predator is at B′ and the prey at C′). In this case the arcs intersect at D′, where the prey arrives after 1.3 time units (from the start of the swerve) and the predator after 1.4 units. The prey crosses the predator's path just ahead of it, and escapes. The predator may turn and continue the chase but it will have lost a lot of ground.

This shows that it is sometimes possible for a prey animal to escape a faster predator, but that it should not swerve too soon. Suppose that the prey delays its swerve to the last possible moment, allowing no margin for error. In this case it will escape if it can accelerate sideways from the line ABC, faster than the predator. An animal running with speed u along a circular arc of radius r has acceleration u^2/r towards the centre of the circle (equation A.6). Therefore the sideways accelerations of the two animals, when they start swerving, are u^2/r for the prey and U^2/R for the predator. The prey can escape only if $u^2/r > U^2/R$. In the case illustrated, $u^2/r = (0.75\,U)^2/0.5R = 1.13\,U^2/R$, so escape is possible.

This analysis is idealized but not wholly unrealistic: gazelles swerve when chased by cheetahs, birds pursued by falcons often swerve upwards, and moths that detect the ultrasonic cries of nearby bats swerve erratically.

For running animals, swerving ability may be limited by the coefficient of friction μ between the feet and the ground. The mean vertical component

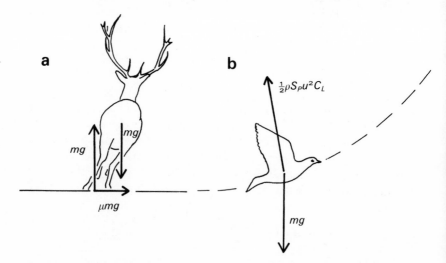

Figure 7.5   Diagrams showing external forces acting on (a) a mammal running on a curve and (b) a bird swerving upwards.

of the force on the feet equals the weight mg of the animal so the mean frictional force cannot exceed μmg (Figure 7.5a) and the greatest possible sideways acceleration is μg. It seems not to be known how the coefficients of friction of the hooves of antelopes compare with those of the paws of predatory mammals. Similar animals of different sizes seem likely to have about equal coefficients of friction, so small running animals may be at no disadvantage compared to large ones, in pursuit.

For flying animals, swerving ability may be limited by the maximum lift that the wings can produce without stalling, which is $\frac{1}{2}\rho S_p \cdot u^2 C_{L\,max}$ (equation A.17). Here ρ is air density, S_p is the plan area of the wings and $C_{L\,max}$ is the maximum attainable lift coefficient. The bird shown in Figure 7.5b is flying at speed u. If it swerves upwards, its greatest possible initial upward acceleration is $(\rho S_p u^2 C_{L\,max}/2m) - g$ or $g[(\rho u^2 C_{L\,max}/2N) - 1]$, where N is the wing loading mg/S_p. Air density ρ is the same for predator and prey, and $C_{L\,max}$ is likely also to be the same. Hence the bird with the greater value of u^2/N should succeed in the chase. Since optimum flying speeds tend to be proportional to $N^{1/2}$ (equation 4.11), predators and prey may often be rather evenly matched.

Predators are often much more likely to succeed, if they take their prey by surprise. This was illustrated by a series of experiments with a goshawk (*Accipiter gentilis*) trained for falconry (Kenward, 1978), which was flown at woodpigeons (*Columba palumbus*) feeding in fields and gardens. On 37 occasions the pigeons were apparently taken by surprise, and did not take off until the hawk was within 20 m of them. In about 50% of these attacks, the hawk caught a pigeon. On another 44 occasions, the pigeons took off when the hawk was more than 20 m away. Only about 5% of these attacks were successful.

Lions and other cats generally stalk their prey so as to get as close as possible before being detected. If they fail to catch their prey quickly, they do not chase it far. In contrast, African hunting dogs (*Lycaon pictus*) often chase gazelles for several kilometres. They hunt in packs of (typically) about a dozen dogs, which gives an important advantage in a chase. The prey may evade the leading dog by swerving when this dog is close behind, but dogs further back in the pack may be well placed to catch it as it swerves.

Herds, flocks and shoals

Many other animals also travel in groups. Antelopes form herds, finches fly in flocks and many fish swim in shoals. These groups are sometimes very large. For instance, shoals of herring (*Clupea harengus*) may be as much as 15 km long and include many millions of individuals.

It is often safer in a group. In the experiments with goshawks described above, almost 80% of attacks on single pigeons were successful but only about 10% of attacks on groups of more than 10 pigeons ended with a pigeon being caught. The reason was probably that when one pigeon detected the hawk and took flight, the rest of the flock were alerted. The whole flock took off almost simultaneously. A hawk is less likely to get close to a flock without being noticed, than to get equally close to a single bird. The same principle probably applies to many other potential prey species, but there is a second reason for safety in numbers (Hamilton, 1971).

Imagine a landscape in which prey animals are scattered, some singly and some in herds. A predator may appear suddenly at any point and kill the nearest prey animal. A particular animal is much less likely to be the victim if it is at the centre of a herd (with close neighbours in every direction) than if it is isolated or at the edge of a herd. Prey animals with an hereditary tendency to push to the centre of any group will be favoured by natural selection, unless the advantage of safety in the group is counteracted by some disadvantage. Prey animals are likely to evolve the tendency to form tight herds, each animal trying to get in to the centre. The herds may make it easier for predators to find prey and so may be to the disadvantage of the prey species as a whole, but non-herding mutants are nevertheless likely to be eliminated by natural selection.

It may sometimes be more economical of energy, as well as safer, to move in a group. The reasons are different for running, swimming and flight.

The air exerts drag on the body of a running animal. This was ignored in Chapter 5 because it accounts for only a small proportion of the energy cost of running, except against strong winds. A small saving of energy may nevertheless be critical in a race. An athlete can reduce the drag on his body by the technique of slipstreaming, that is by keeping close behind another runner. At 6 m/s (a middle-distance speed), he can reduce his energy consumption in this way by up to 6.5% in still air. This has been demonstrated by measuring the oxygen consumption of men running singly or one behind another on a treadmill in a wind tunnel. (A wind blowing at the speed of the treadmill was required, in this situation, to simulate running on stationary ground in still air). Though the tactic of slipstreaming is used by athletes, I know no examples of its use by other animals.

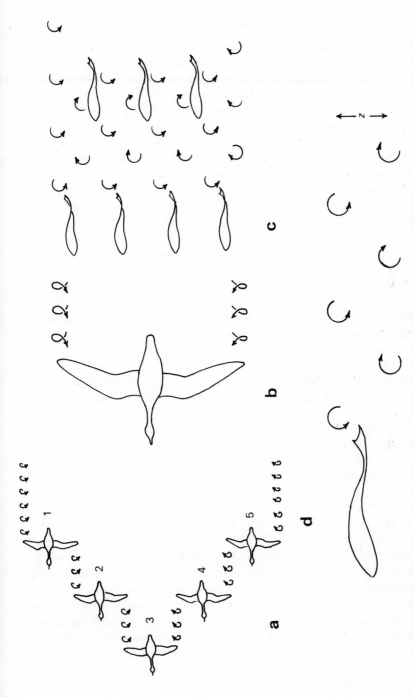

Figure 7.6 Diagrammatic plan views of (a) a flock of geese in flight, (b) a single goose in flight, (c) fish swimming in two rows and (d) a single fish swimming. Vortices are indicated by curved arrows.

Birds may be able to save energy by flying in flocks, especially in the V-formations used by geese (Figure 7.6a—Lissaman and Shollenberger, 1970). Aircraft travel most economically at the speed (u_3) that minimizes drag (p. 69). At this speed, induced drag is half of the total drag. Induced drag is least for aircraft with high aspect ratios (long narrow wings), which suggests that a group of aircraft could save fuel by flying in formation, side by side with wing tips touching, so as to resemble a single aircraft of much higher aspect ratio. A similar effect could be obtained without wings touching, if the aircraft flew in a V-formation like geese.

To understand how this would work, consider the air movements behind the wings of an aircraft or of a gliding bird (Figure 7.6b). The air passing immediately over and under the wings is deflected downwards (as it must be, to produce lift) but the air beyond the wing tips is not. Consequently, vortices form behind the wing tips. The work done against induced drag provides the kinetic energy of these vortices.

Notice that the vortices behind the left and right wing tips, in Figure 7.6b, have opposite directions of rotation. In Figure 7.6a, the left wing tip of bird number 2 is in the vortex from the right wing tip of bird number 3. Since it tends to produce a vortex of opposite direction, its effect is to cancel out the first vortex. The only vortices that are not cancelled out in this way are those behind the right wing of bird number 1 and the left wing of bird number 5. Much less kinetic energy is left in the wake than if the five birds were flying separately, and induced drag is correspondingly reduced.

This simple account of flight in V-formation has ignored various complications, including the flapping of the wings, but it indicates a principle whereby energy may be saved.

A similar suggestion has been made about shoaling fish (Weihs, 1973). The tail of a fish, driving water alternately to left and right (Figure 2.5b) leaves behind it two lines of vortices in which the water is rotating in opposite directions (Figure 7.6d). Let the distance between the two lines be z, and imagine two rows of fish with the fish in each row $2z$ apart (Figure 7.6c). The fish in the second row could take up positions as shown, so that they tended to produce clockwise vortices where anticlockwise ones already existed, and vice versa. The second row would cancel out the vortices produced by the first so less kinetic energy would be left behind in the wake than if the fish were swimming separately. This suggests that energy might be saved by fish swimming in staggered rows, each fish $2z$ from his neighbours on either side. Unfortunately for the theory, fish swimming in a large tank failed to space themselves as predicted (Partridge and Pitcher, 1979). Videotape recordings showed for instance that saithe (*Pollachius virens*) tend to keep $6z$ from their neighbours on either side.

This mechanism could in theory save energy for the second row of fish in Figure 7.6c, but not for the first. It could save energy only for alternate rows of fish in a shoal. The habit of using this mechanism seems unlikely to evolve except perhaps in species that form shoals of close relatives, because the savings made by some members of a shoal would depend on the cooperation of others who derived no benefit.

A slightly different mechanism could save energy if alternate fish in each row swam out of phase with each other, so that a fish beat its tail to the left while its neighbours beat theirs to the right. The movements of neighbouring fish would be tending to produce opposite vortices in the same place and so (ideally) no vortices would be formed. Fish probably obtain little or no benefit from this mechanism. It would work best with neighbouring fish only z apart, far closer than fish actually swim.

APPENDIX

BASIC MECHANICS

This appendix explains the basic mechanics used in the rest of the book. Explanations of things likely to have been covered at school are very brief, and intended merely as revision.

Units

S.I. (Système Internationale) units are used. The system is based on the metre (m), kilogram (kg), second (s), kelvin (K) and a few other units that are not needed here.

The kelvin is the unit of temperature. It is the same size as the degree centigrade but a temperature of 0 K is the absolute zero, and 0°C is 273 K. Angles are measured in radians (rad, about 57°). Figure A.1a shows how

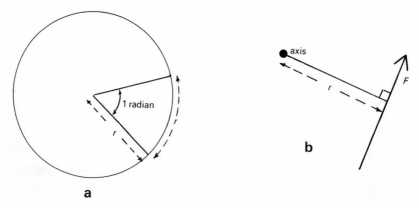

a

b

Figure A.1 (a) An angle of one radian; (b) the moment (Fr) of a force F acting at a distance r from the axis.

the radian is defined. Frequencies are measured in herz (Hz, cycles per second). The following additional units are defined later:

the *newton* (N) for force   the *joule* (J) for energy
the *watt* (W) for power   the *pascal* (Pa) for stress

Prefixes are used to indicate fractions or multiples of units.

milli- (m-) for 10^{-3} kilo- (k-) for 10^3
micro- (μ-) for 10^{-6} mega- (M-) for 10^6
nano- (n-) for 10^{-9} giga- (G-) for 10^9

Thus 1 μm means 10^{-6} metres and 1 GPa means 10^9 pascals.

Vectors

Some quantities such as mass can be described fully by a number and a unit (for instance, 7 kg). They are called *scalars*. Some other quantities such as force and velocity have direction as well as size and are not fully described unless the direction is stated (for instance, a velocity of 5 m/s due north). They are called *vectors*.

A vector v acting at an angle θ to the horizontal can be resolved into a horizontal component $v \cos \theta$ and a vertical component $v \sin \theta$ (Figure A.2a). The two components, acting together, have exactly the same effect as the original vector, which is described as the *resultant* of the two components. Similarly, components x horizontally and y vertically have a resultant $(x^2 + y^2)^{1/2}$ at an angle arctan y/x to the horizontal (Figure A.2b).

Movement

Consider a particle of matter moving in a straight line. Its position, shown by a scale marked along the line, is 0 at time 0 and x at time t (Figure A.2c). Its velocity u at time t is given by

$$u = dx/dt \qquad \qquad \text{A.1}$$

Its acceleration a is the rate of change of velocity

$$a = du/dt = d^2x/dt^2 \qquad \qquad \text{A.2}$$

One of the laws of motion given in the next section states that the acceleration of a body is constant if the forces on it are constant. Let the body being considered have constant acceleration. Then

$$u = u_0 + at \qquad \qquad \text{A.3}$$

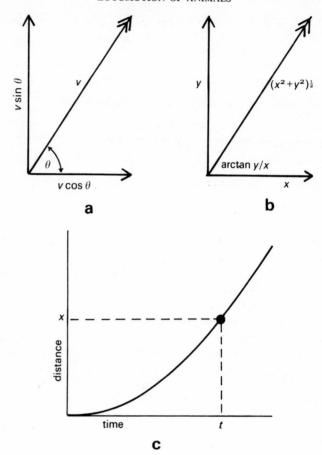

Figure A.2 (a), (b) Resultants of vectors; (c) distance vs. time graph.

where u_0 is the velocity at time 0. This equation says that the increase of velocity is (acceleration) × (time). Also

$$x = (u_0 + u)t/2 \qquad\qquad \text{A.4}$$

This equation says that distance travelled is (mean velocity) × (time). Further

$$x = u_0 t + \tfrac{1}{2}at^2 \qquad\qquad \text{A.5}$$

Remember that equations A.3 to A.5 apply only if the acceleration is constant.

A body moving in a circle of radius r with constant speed u does not have constant velocity, because its direction of movement is always changing. It has acceleration u^2/r towards the centre of the circle.

$$a = u^2/r \qquad\qquad \text{A.6}$$

Notice that in this case the acceleration is at right angles to the direction of movement.

A rotating body that has turned through angle θ in time t has angular velocity $d\theta/dt$ and angular acceleration $d^2\theta/dt^2$.

Force

The concept of force is explained by Newton's three Laws of Motion.

(i) A body remains at rest or moves at constant velocity unless forces act on it.

(ii) An unbalanced force F acting on a body of mass m gives the body an acceleration a in the direction of the force, where

$$F = m \cdot a = m \cdot du/dt \qquad\qquad \text{A.7}$$

(see equation A.2). The newton is defined as the force that gives a mass of 1 kg an acceleration of 1 m/s^2.

(iii) If body A exerts a force on body B, body B exerts an equal, opposite force on body A.

The *momentum* of a body is its mass multiplied by its velocity, *mu*. Newton's second law can therefore be expressed by saying that the rate of change of momentum of a body $(m \cdot du/dt)$ is equal to the force acting on it. From this follows the Law of Conservation of Momentum: the total momentum of a system remains constant unless external forces act on it.

The *weight* of a body is the force exerted on it by gravity. A body falling freely in a vacuum has a downward acceleration g (9.8 m/s^2 on earth) so a body of mass m has weight mg. Since weight is a force it is measured in newtons. Though every part of the body has weight it is generally convenient to think of the weight as a single force, acting at the centre of mass of the body.

In discussion of rotation, moments take the place of forces and moments of inertia take the place of masses. The *moment* that a force exerts about an axis is the force multiplied by the perpendicular distance of the line of action of the force from the axis. In Figure A.1b, the force F is in the plane of the paper and the axis is at right angles to the paper. The force exerts a moment Fr about the axis. *Moments of inertia* tell us about the moments needed to give angular accelerations to bodies. Consider a body pivoted at an axis about which it has moment of inertia I. A moment T acting about the same axis gives it an angular acceleration $d^2\theta/dt^2$ where

$$T = I \cdot d^2\theta/dt^2 \qquad\qquad\qquad \text{A.8}$$

A body that is in *equilibrium* has neither linear nor angular acceleration. Consider a body acted on by forces which all act in the same vertical plane. It is in equilibrium if

(i) the total of the vertical components of all forces is zero (i.e. the downward forces balance the upward forces), and

(ii) the total of the horizontal components of all the forces is zero (i.e. forces to the left balance forces to the right) and

(iii) the total of the moments of the forces about a chosen point in the plane is zero (i.e. anticlockwise moments balance clockwise moments).

Energy

Work is done when forces move their points of application. If the point of application moves a distance x and the component of the force in the direction of movement is F_X, the work done is

$$W = F_X \cdot x \qquad\qquad\qquad \text{A.9}$$

If F_X is one newton and x is one metre the work is one joule. Notice that this equation makes the work negative if the force and movement are in opposite directions. A muscle that shortens while exerting tension does positive work but one that is forcibly extended while exerting tension does negative work (i.e. positive work is done on it).

If a body of mass m, weight mg is lifted through a height h, work mgh is done on it and its gravitational potential energy is increased by mgh. A body of mass m moving with velocity u has kinetic energy $\frac{1}{2}mu^2$. Similarly a body that has moment of inertia I about an axis, and is rotating with angular velocity $d\theta/dt$ about the same axis, has kinetic energy $\frac{1}{2}I(d\theta/dt)^2$. Potential and kinetic energy (and all other forms of energy) are measured in joules, like work. Positive work has to be done on a body to increase its (potential + kinetic) energy, and negative work has to be done to decrease this energy. My muscles do positive work (increasing my potential energy) as I climb up stairs, and negative work as I come down.

Power is the rate of doing work. If one joule is done per second the power is one watt.

Energy cannot be created or destroyed (except in atomic explosions, etc.) so no process involves a net loss of energy. A body falling in a vacuum is gaining kinetic energy as fast as it loses potential energy. In many processes, however, energy is wasted by conversion to an unusable form. For instance, a muscle using chemical energy (from foodstuffs) to do work wastes a large proportion of the energy as heat. The efficiency of any machine is the ratio (output of useful work)/(input of energy).

The *heat of combustion* of a substance is the heat released when unit mass of the substance is burned completely. The heats of combustion of fats and carbohydrates are about 40 and 17 MJ/kg, respectively. Metabolism using 1 cm^3 oxygen releases about 20 J chemical energy, whatever food is being oxidized.

Properties of materials

The *density* of a substance is the mass of unit volume. The densities of pure liquid water and of sea water at ordinary temperatures are about 1000 and 1026 kg/m^3, respectively. The density of air is considerably affected by changes of temperature and pressure; it is 1.2 kg/m^3 at 20°C and 1 atmosphere pressure.

Pressure is measured in pascals, which are units of one newton per square metre. One atmosphere is 100 kPa. The pressure P at depth h in a liquid of density ρ is

$$P = P_0 + \rho g h \qquad\qquad \text{A.10}$$

where P_0 is the pressure at the surface. Boyle's Law states that if the pressure on a quantity of gas changes from P_1 to P_2 and the volume from V_1 to V_2 while the temperature remains constant

$$P_1 V_1 = P_2 V_2 \qquad\qquad \text{A.11}$$

Stress, like pressure, is force per unit area and is measured in pascals. Figure A.3a shows a block of material of cross-sectional area A. Forces F pulling on its ends (Figure A.3b) set up a tensile stress F/A in it; forces F pushing on its ends (Figure A.3c) set up a compressive stress F/A, and forces F acting on the ends as shown in Figure A.3d set up a shear stress F/A.

Strain is deformation under stress. In Figure A.3b the block has been stretched from its initial length x_0 to a new length x. It has suffered a tensile strain $(x - x_0)/x_0$. In Figure A.2d the block has suffered a shear strain y/x_0.

The strain in an elastic solid is related to the stress. If the solid obeys Hooke's Law, strain is proportional to stress. For tensile strain

$$F/A = E(x - x_0)/x \qquad \text{A.12}$$

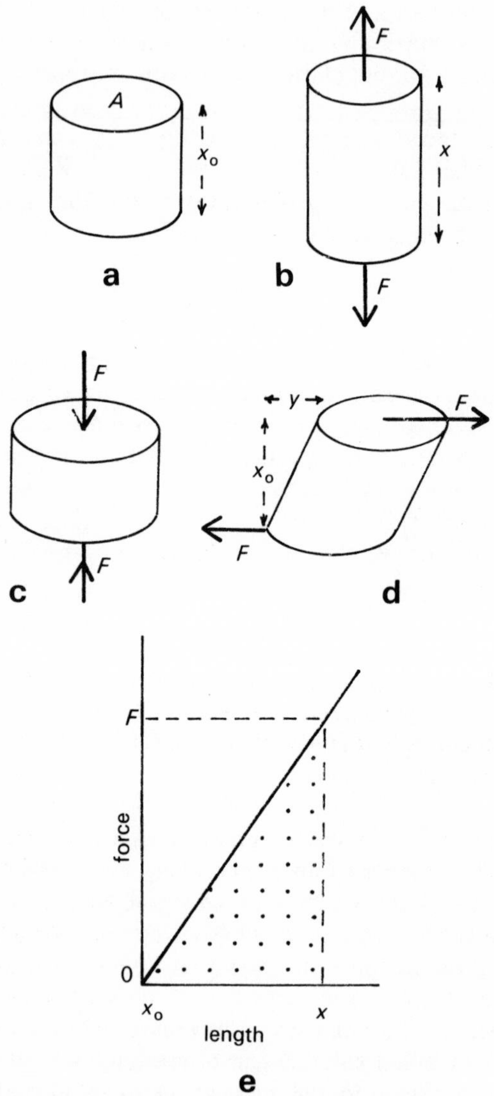

Figure A.3 Diagrams used to explain stress and elastic strain energy.

The constant E is called *Young's modulus*. It is (tensile stress)/(tensile strain), as the equation shows.

When the stress on an elastic body is removed there is an elastic recoil and the strain vanishes. A force distorting a block of elastic material does work on it and stores in it elastic strain energy, which can be recovered in an elastic recoil. If the block shown in Figure A.3b obeys Hooke's Law, a graph of tensile force F against extension $(x - x_0)$ is a straight line (Figure A.3e). The strain energy stored in it is the area under the graph, or $F(x - x_0)/2$.

Viscosity is the property that makes treacle (molasses) hard to stir. The stress in a viscous liquid is not related to the strain but to the rate at which the strain is changing. Ideally, it is proportional to the strain rate$(dy/dt)/x_0$ in Figure A.3d.

$$F/A = \eta (dy/dt)/x_0 \qquad\qquad \text{A.13}$$

The constant η is called the viscosity. There is no elastic recoil when the stress is removed from a viscous liquid.

Friction resists sliding of solid surfaces over each other. The frictional force F (Figure 5.12b,d) may have any value up to a maximum F_{max} which is roughly proportional to the normal force N. The ratio F_{max}/N is called the *coefficient of friction*.

Similarity

Two triangles are geometrically similar if they have the same angles, so that the smaller could be made to match the larger by magnifying it. More generally, two shapes are geometrically similar if one can be made to match the other by a uniform change of scale. Geometrically similar bodies have volumes proportional to (length)^3 and areas proportional to (length)^2.

Two motions of similar systems are dynamically similar if one could be made to match the other by uniform changes of one or both of the scales of length and time. For instance, the motions of two pendulums of different length swinging through the same angle are dynamically similar.

When gravitational forces are important (for instance, in the swinging of pendulums) dynamic similarity is only possible if the motions have equal *Froude numbers*, u^2/gl. Here u is velocity, measured at corresponding stages of the two motions. In the case of pendulums, for instance, it might be the velocity at the bottom of the swing. Also l is a length, taking corresponding lengths for the two motions. In the case of pendulums the obvious choice would be the overall length from support to bob. Finally, g is the acceleration of free fall.

When viscous forces are important (for instance, in the flow of fluids around solid bodies) dynamic similarity is only possible if the motions have equal *Reynolds numbers*, $\rho u l / \eta$. Here u and l are again a velocity and a length, and ρ and η are the density and viscosity of the fluid. For water, $\rho / \eta = 10^6 \, \text{s m}^{-2}$ so Reynolds number is 10^6 (length in m) (velocity in m/s). For air, $\rho / \eta = 7 \times 10^4 \, \text{s m}^{-2}$.

It does not matter exactly how the lengths and velocities are defined, for calculating Reynolds and Froude numbers, provided that consistent definitions are used. They should, however, be reasonably typical of the motion: it would be silly to use the diameter of the string as the characteristic length in discussion of pendulums. In discussions of bodies moving through fluids, the overall length and mean velocity are generally used.

Reynolds and Froude numbers are dimensionless, so they have no units. The discussion of Reynolds numbers continues in the next section.

Bodies in fluids

A body of mass m and volume V has weight mg. The weight is a downward force. Archimedes' Principle states that if the body is submerged in a fluid of density ρ an upward force $\rho V g$ also acts on it. Just as weight can conveniently be considered to act at the centre of mass, the upthrust can be considered to act at the centre of buoyancy, which is the geometrical centre of the body.

The forces on bodies moving through fluids depend on the patterns of flow of the fluid around them, and so on the Reynolds number. Figure A.4 shows changes in the pattern of flow that occur as Reynolds number increases. In every case the body drags fluid along with it. The fluid immediately in contact with the body has the same velocity as the body and tends to drag more distant fluid along too, because of viscosity. Forces act on the body because of the viscosity of the fluid and also because of the inertia of the fluid (i.e. because forces are needed to accelerate it). Viscous forces tend to be dominant at low Reynolds numbers and inertia forces at high ones.

At very low Reynolds numbers (Figure A.4a) fluid velocity falls off very gradually with distance from the body, in every direction. Nevertheless, the fluid is at rest far behind the body: the body leaves no wake of disturbed fluid. At higher Reynolds numbers (Figure A.4b) fluid velocity falls off much more steeply so that fluid movement alongside the body is

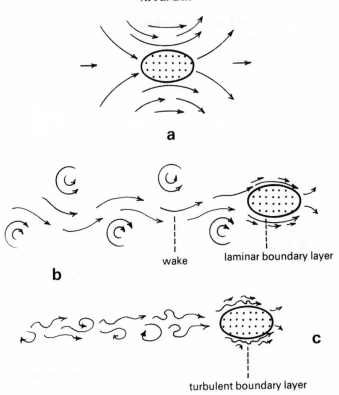

Figure A.4 Diagrams of a body moving to the right through a fluid. The arrows show movement of the fluid caused by the passage of the body. The lengths of the arrows indicate velocities. The Reynolds numbers are (a), less than 1; (b), between 100 and about 10^6 and (c), greater than about 10^6.

virtually restricted to a thin "boundary layer". The body leaves a wake of moving, swirling fluid behind itself. At moderate Reynolds numbers (Figure A.4b) flow in the boundary layer is smooth, everywhere parallel to the surface of the body (laminar flow). At high Reynolds numbers (Figure A.4c) there is irregular, swirling flow (turbulent flow) in the boundary layer. The change from flow pattern (a) to (b) (Figure A.4) occurs gradually over a wide band of Reynolds numbers. The change from (b) to (c) occurs abruptly, at slightly different Reynolds numbers for bodies of different shapes.

The fluid resists the movement of the body. It exerts on the body a backward component of force, called *drag*. At Reynolds numbers less than

1 the drag on a body of length l moving with velocity u through fluid of viscosity η can be calculated from the equation

$$\text{drag} = k\eta lu \qquad \text{A.14}$$

where k is a constant that depends on the shape of the body. (It is 3π for a sphere). At Reynolds numbers greater than 100 the drag can be calculated from the equation

$$\text{drag} = \tfrac{1}{2}\rho Su^2 C_D \qquad \text{A.15}$$

where S is an area and C_D is a dimensionless number called the *drag coefficient*. Confusingly, several different areas may be used in this equation. The wetted area S_w is the total surface area of the body. The frontal area S_f is the area of a full-scale front view of the body. The plan area S_p is the area of a full-scale plan of the body and is the area usually used in calculations about aerofoils and hydrofoils. The value of the drag coefficient depends on which area is used.

The drag described by equation A.15 has two components, friction drag due to the viscosity of the boundary layer and pressure drag due to higher pressure in front of the body than behind it. It can generally be assumed that in the range of Reynolds numbers that give a laminar boundary layer

$$\text{friction drag} \simeq \tfrac{1}{2}\rho S_w u^2 (1.3R^{-1/2}) \qquad \text{A.16}$$

where R is the Reynolds number. This implies that the friction drag coefficient based on wetted area is $1.3R^{-1/2}$, that is about 0.04 at a Reynolds number of 10^3 and 0.004 at a Reynolds number of 10^5. Friction drag is the principal component of drag for flat plates moving edge-on, and also for well streamlined (torpedo-shaped) bodies which are designed to leave as little disturbance as possible in their wake.

The pressure drag coefficient depends on the shape of the body and also on its attitude. (It is different for a cylinder moving end-on from one moving broadside-on). For bodies of given shape and attitude it is generally fairly constant over a wide range of Reynolds numbers, from about 1000 to about 10^5. For instance, the pressure drag coefficient based on wetted area for spheres in this range is about 0.12. Spheres and other unstreamlined bodies leave a lot of disturbance in their wake, and the pressure drag on them is much larger than the friction drag.

Drag acts directly backwards along the direction of motion. When a symmetrical body moves through a fluid, parallel to its axis of symmetry, drag is the only component of force that the fluid exerts on it. If the body is asymmetrical or moves at an angle to its axis of symmetry there is

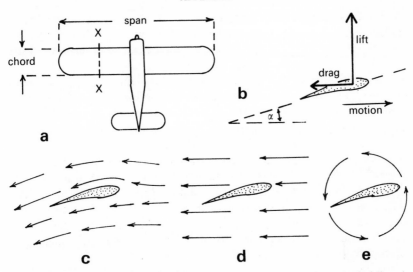

Figure A.5 (a) Diagram of an aeroplane; (b) section through an aerofoil or hydrofoil, such as a vertical section along XX in (a). Lift, drag and the angle of attack (α) are shown. (c) Section through a hydrofoil that is producing lift. Fluid flow relative to the hydrofoil is indicated by arrows. (d), (e) Flow around a hydrofoil as a combination of uniform flow and circulation around the hydrofoil.

another component of force called *lift*, acting at right angles to its direction of motion. This is the force that prevents aeroplanes from falling. Aerofoils and hydrofoils are structures such as aeroplane wings and propeller blades, designed to give maximum lift for minimum drag. The lift and drag on a hydrofoil depend on its angle of attack (α, Figure A.5b) as well as on its speed. Lift is zero at some small angle of attack and increases as the angle is increased until it reaches a maximum at an angle of attack of (typically) about 20°. At greater angles of attack the lift is less, because the fluid then flows irregularly over the hydrofoil. (This phenomenon is called *stalling*.) The drag has a minimum value at some small angle of attack and increases as the angle of attack increases. For any particular angle of attack, a lift coefficient C_L and a drag coefficient C_D (both based on plan area) are defined by the equations

$$\text{lift} = \tfrac{1}{2}\rho S_p u^2 C_L \qquad\qquad \text{A.17}$$
$$\text{drag} = \tfrac{1}{2}\rho S_p u^2 C_D \qquad\qquad \text{A.18}$$

Reynolds numbers are calculated for hydrofoils, using the chord (Figure A.5a) as the length. Well-designed hydrofoils give maximum lift coefficients of about 1.0 at Reynolds numbers around 10^3 and 1.5 at around 10^6. A hydrofoil designed for one range of Reynolds numbers is not necessarily suitable for another. For instance, an enlarged model of an insect wing would not be suitable for an aeroplane.

Hydrofoils produce lift by accelerating air at right angles to their path (Figure A.5c). For instance, an aeroplane wing drives air downwards. It exerts a downward force on the air and the air exerts an equal, upward force on it. By Newton's second Law of Motion, the lift is equal to the rate at which the air is given downward momentum.

In an earlier paragraph, drag was broken down into friction drag and pressure drag. In discussions of hydrofoils it is convenient to divide drag differently, into profile drag and induced drag. Profile drag would act even if the angle of attack were adjusted so that there was no lift. Induced drag acts because lift is being produced: the work done against induced drag equals the kinetic energy given to the fluid being accelerated at right angles to the hydrofoil's path.

$$\text{drag} = \text{profile drag} + \text{induced drag}$$
$$\simeq \tfrac{1}{2}\rho S_p u^2 C_{D0} + (L^2/2\rho S_p u^2 A) \qquad \text{A.19}$$

Here C_{D0} is a drag coefficient that takes account of profile drag only, L is the lift and A is the aspect ratio (span/chord, Figure A.5b). Notice that high aspect ratios give low induced drag. The expression given for induced drag in equation A.19 has been simplified by assuming a typical value for a factor, usually given the symbol k, which depends on the shape of the hydrofoil.

The mechanism by which lift is produced depends on fluid flowing faster over one surface of the hydrofoil than the other. (In Figure A.5c, flow is faster where the arrows are longer). It is often convenient to think of the flow past a hydrofoil as a combination of uniform flow with the same velocity everywhere (Figure A.5d), together with circulation around the hydrofoil (Figure A.5e).

Helicopters support their weight by driving a column of air downwards (Figure A.6a). This requires power equal to the rate at which kinetic energy is given to the air. For a helicopter of weight mg, with a rotor of radius r, hovering in air of density ρ, this power is given by

$$\text{induced power} = (m^3 g^3/2\rho\pi r^2)^{1/2} \qquad \text{A.20}$$

This is the power needed to overcome induced drag on the rotor blades, and additional power is needed to overcome profile drag.

A helicopter rotor produces a continuous column of moving air but

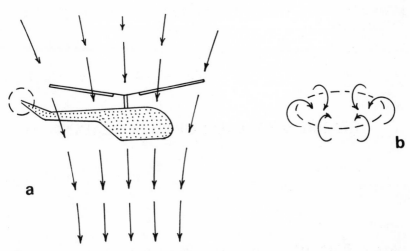

Figure A.6 Diagrams of (a) a helicopter and (b) a vortex ring. Fluid flow is indicated by arrows.

many animal movements produce a series of puffs of air or of moving water. Such puffs of moving fluid form rotating rings, called vortex rings (Figure A.6b) in which the fluid is rotating about the circular core which is itself moving at right angles to its plane. Smoke rings are an everyday example of vortex rings.

REFERENCES AND FURTHER READING

General

Alexander, R. McN. and Goldspink, G. (eds.) (1977) *Mechanics and Energetics of Animal Locomotion*. Chapman & Hall, London.
Day, M. H. (ed.) (1981) Vertebrate locomotion. *Symp. zool. Soc.* **48**, in the press.
Elder, H. Y. and Trueman, E. R. (eds.) (1980) *Aspects of Animal Movement*. Cambridge University Press, Cambridge.
Gray, J. (1968) *Animal Locomotion*. Weidenfeld & Nicolson, London.
Nachtigall, W. (ed.) (1977) Physiology of movement—biomechanics. *Fortschr. Zool.* **24** (2–3), 1–352.
Pedley, T. J. (ed.) (1977) *Scale Effects in Animal Locomotion*. Academic Press, London.

Chapter 1

Alexander, R. McN. (1969) The orientation of muscle fibres in the myomeres of fish. *J. mar. biol. Ass. U.K.* **49**, 263–290.
Armstrong, R. B. (1981) Recruitment of muscles and fibres within muscles in running animals, in Day (1981) (*see* General).
Bone, Q. (1966) On the function of the two types of myotomal muscle fibre in elasmobranch fish. *J. mar. biol. Ass. U.K.* **46**, 321–349.
Clark, R. B. (1964) *Dynamics in Metazoan Evolution*. Clarendon, Oxford.
Currey, J. D. (1980) Skeletal factors in locomotion, in Elder and Trueman (1980) (*see* General) 27–48.
Holberton, D. V. (1977) Locomotion of Protozoa and single cells, in Alexander and Goldspink (1977) (*see* General) 279–332.
Margaria, R. (1976) *Biomechanics and Energetics of Muscular Exercise*. Clarendon Press, Oxford.
Ruben, J. A. and Bennett, A. F. (1980) Antiquity of the vertebrate pattern of activity metabolism and its possible relation to vertebrate origins. *Nature* **286**, 886–888.
Usherwood, P. N. R. (ed.) (1975) *Insect Muscle*. Academic Press, New York.
Ward, D. V. (1972) Locomotory function of squid mantle. *J. Zool., Lond.* **167**, 437–449.
White, D. C. S. (1977) Muscle mechanics, in Alexander and Goldspink (1977) (*see* General) 23–56.

Chapter 2

Blake, J. R. and Sleigh, M. A. (1974) Mechanics of ciliary locomotion. *Biol. Rev.* **49**, 85–125.

Blake, R. W. (1978) On balistiform locomotion. *J. mar. biol. Ass. U.K.* **58**, 73–80.

Clark, R. B. and Tritton, D. J. (1970) Swimming mechanisms in nereidiform polychaetes. *J. Zool., Lond.* **161**, 257–271.

Clarke, B. D. and Bemis, W. (1979) Kinematics of swimming of penguins at the Detroit Zoo. *J. Zool. Lond.* **188**, 411–428.

Gladfelter, W. B. (1972) Structure and function of the locomotory system of *Polyorchis montereyensis* (Cnidaria, Hydrozoa). *Helgoländer wiss. Meeresunters.* **23**, 38–79.

Hoar, W. S. and Randall, D. J. (eds.) (1978) *Fish Physiology 7: Locomotion.* Academic Press, New York.

Lighthill, M. J. (1969) Hydromechanics of aquatic animal propulsion. *Ann. Rev. Fluid Mech.* **1**, 413–446.

McCutchen, C. W. (1977) Froude propulsive efficiency of a small fish, measured by wake visualisation, in Pedley (1977) (*see* General) 339–363.

Machemer, H. (1972) Ciliary activity and the origin of metachrony in *Paramecium*: effects of increased viscosity. *J. exp. Biol.* **57**, 239–259.

Nachtigall, W. (1980) Mechanics of swimming in water-beetles, in Elder and Trueman (1980) (*see* General) 107–124.

Prange, H. D. and Schmidt-Nielsen, K. (1970) The metabolic cost of swimming in ducks. *J. exp. Biol.* **53**, 763–778.

Sleigh, M. A. and Blake, J. R. (1977) Methods of ciliary propulsion and their size limitations, in Pedley (1977) (*see* General) 243–256.

Trueman, E. R. (1980) Swimming by jet propulsion, in Elder and Trueman (1980) (*see* General) 93–105.

Wardle, C. S. and Videler, J. J. (1980) Fish swimming, in Elder and Trueman (1980) (*see* General) 125–150.

Webb, P. W. (1971) The swimming energetics of trout. *J. Exp. Biol.* **55**, 489–540 (two papers).

Webb, P. W. (1975) Hydrodynamics and energetics of fish propulsion. *Bull. Fish. Res. Bd. Can.* **190**, 1–158.

Wu, T. Y. (1977) Introduction to the scaling of aquatic animal locomotion, in Pedley (1977) (*see* General) 203–232.

Chapter 3

Alexander, R. McN. (1965) The lift produced by the heterocercal tails of Selachii. *J. exp. Biol.* **43**, 131–138.

Alexander, R. McN. (1966) Physical aspects of swimbladder function. *Biol. Rev.* **41**, 141–176.

Alexander, R. McN. (1972) The energetics of vertical migration by fishes. *Symp. Soc. exp. Biol.* **26**, 273–294.

Bidigare, R. R. and Biggs, D. C. (1980) The role of sulfate exclusion in buoyancy maintenance by siphonophores and other gelatinous zooplankton. *Comp. Biochem. Physiol.* **66A**, 467–471.

Blaxter, J. H. S. and Tytler, P. (1978) Physiology and function of the swimbladder. *Adv. comp. Physiol. Biochem.* **7**, 311–367.

Corner, E. D. S., Denton, E. J. and Forster, G. R. (1969) On the buoyancy of some deep-sea sharks. *Proc. R. Soc. B.* **171**, 415–429.

Denton, E. J. (1974) On buoyancy and the lives of modern and fossil cephalopods. *Proc. R. Soc. B.* **185**, 273–299.

Magnusson, J. J. (1978) Locomotion by scombroid fishes: hydromechanics, morphology and behaviour, in *Fish Physiology* (ed. W. S. Hoar and D. J. Randall) **7**, 239–313.

Simons, J. R. (1970) The direction of the thrust produced by the heterocercal tails of two dissimilar elasmobranchs: the Port Jackson shark, *Heterodontus portusjacksoni* (Meyer) and the piked dogfish, *Squalus megalops* (Macleay). *J. exp. Biol.* **52**, 95–107.

Chapter 4

Baker, P. S. and Cooter, R. J. (1979) The natural flight of the migratory locust, *Locusta migratoria* L. *J. comp. Physiol.* A **131**, 79–87 (two papers).
Cloupeau, M., Devillers, J. F. and Devezeaux, D. (1979) Direct measurements of instantaneous lift in desert locust; comparison with Jensen's experiments on detached wings. *J. exp. Biol.* **80**, 1–15.
Dathe, H. H. and Oehme, H. (1978) Typendes Rüttelfluges der Vögel. *Biol. Zbl.* **97**, 299–306.
Ellington, C. P. (1978) The aerodynamics of normal hovering flight: three approaches, in *Comparative Physiology—Water, Ions and Fluid Mechanics* (ed. K. Schmidt-Nielsen, L. Bolis and S. H. P. Maddrell) 327–345. Cambridge University Press, Cambridge.
Gibo, D. L. and Pallett, M. J. (1979) Soaring flight of monarch butterflies, *Danaus plexippus* (Lepidoptera: Danaidae), during the late summer migration in southern Ontario. *Can. J. Zool.* **57**, 1393–1401.
Kokshaysky, N. V. (1979) Tracing the wake of a flying bird. *Nature* **279**, 146–148.
Lighthill, J. (1977) Introduction to the scaling of aerial locomotion, in Pedley (1977) (*see* General) 365–404.
McGahan, J. (1973) Flapping flight of the Andean condor in nature. *J. exp. Biol.* **58**, 239–253.
Nachtigall, W. (1979) Gleitflug des Flugbeutlers, *Petaurus breviceps papuanus* II. Filmanalysen zur Einstellung von Gleitbahn und Rumpf sowie zur Steuerung des Gleitflugs. *J. comp. Physiol.* **133**, 89–95.
Norberg, U. M. (1976) Some advanced flight manoeuvres of bats. *J. exp. Biol.* **64**, 489–495.
Pennycuick, C. J. (1968) A wind-tunnel study of gliding flight in the pigeon *Columba livia. J. exp. Biol.* **49**, 509–526.
Pennycuick, C. J. (1971) Control of gliding angle in Rüppell's griffon vulture, *Gyps ruppellii. J. exp. Biol.* **55**, 39–46.
Pennycuick, C. J. (1972) Soaring behaviour and performance of some East African birds, observed from a motor-glider. *Ibis* **114**, 178–218.
Pennycuick, C. J. (1975) Mechanics of flight, in *Avian Biology* (ed. D. S. Farner and J. R. King) **5**, 1–75, Academic Press, New York.
Pennycuick, C. J. and Lock, A. (1976) Elastic energy storage in primary feather shafts. *J. exp. Biol.* **64**, 677–689.
Rayner, J. M. V. (1979) A new approach to animal flight mechanics. *J. exp. Biol.* **80**, 17–54.
Rüppell, G. (1977) *Bird Flight*. Van Nostrand Reinhold, New York.
Savage, S. B., Newman, B. G. and Wong, D. T.-M. (1979) The role of vortices and unsteady effects during the hovering flight of dragonflies. *J. exp. Biol.* **83**, 59–77.
Tucker, V. A. (1968) Respiratory exchange and evaporative water loss in the flying budgerigar. *J. exp. Biol.* **48**, 67–87.
Weis-Fogh, T. (1973) Quick estimates of flight fitness in hovering animals, including novel mechanisms for lift production. *J. exp. Biol.* **59**, 169–230.
Wilson, J. A. (1975) Sweeping flight and soaring by albatrosses. *Nature* **257**, 307–308.
Withers, P. C. and Timko, P. L. (1977) The significance of ground effect to the aerodynamic cost of flight and energetics of the black skimmer (*Rhyncops nigra*). *J. exp. Biol.* **70**, 13–26.

Chapter 5

Alexander, R. McN. (1976) Estimates of speeds of dinosaurs. *Nature* **261**, 129–130.
Alexander, R. McN. (1980) Optimum walking techniques for quadrupeds and bipeds. *J. Zool., Lond.* **192**, 97–117.

Alexander, R. McN. (1981) Tetrapod gaits: adaptations for stability and economy. *Symp. zool. Soc.* **48** (in the press).

Alexander, R. McN., Jayes, A. S. and Ker, R. F. (1980) Estimates of energy cost for quadrupedal running gaits. *J. Zool., Lond.* **190**, 155–192.

Alexander, R. McN. and Vernon, A. (1975) Mechanics of hopping by kangaroos (Macropodidae). *J. Zool., Lond.* **177**, 265–303.

Bennet-Clark, H. C. (1975) The energetics of the jump of the locust *Schistocerca gregaria. J. exp. Biol.* **63**, 53–83.

Bennet-Clark, H. C. and Alder, G. M. (1979) The effect of air resistance on the jumping performance of insects. *J. exp. Biol.* **82**, 105–121.

Bennet-Clark, H. C. and Lucey, E. C. A. (1967) The jump of the flea: a study of the energetics and a model of the mechanism. *J. exp. Biol.* **47**, 59–76.

Cartmill, M. (1979) The volar skin of primates: its frictional characteristics and their functional significance. *Am. J. phys. Anthropol.* **50**, 497–510.

Cavagna, G. A., Heglund, N. C. and Taylor, C. R. (1977) Mechanical work in terrestrial locomotion: two basic mechanisms for minimizing energy expenditure. *Am. J. Physiol.* **233**, R243–R261.

Delcomyn, F. (1971) The locomotion of the cockroach *Periplaneta americana. J. exp. Biol.* **54**, 443–452.

Gambaryan, P. P. (1974) *How Mammals Run.* Wiley, New York.

Hildebrand, M. (1976) Analysis of tetrapod gaits: general considerations and symmetrical gaits, in *Neural Control of Locomotion* (ed. R. M. Herman, S. Grillner, P. S. G. Stein and D. G. Stuart) Plenum, New York, 203–236.

Hildebrand, M. (1977) Analysis of asymmetrical gaits. *J. Mammal.* **58**, 131–156.

Jayes, A. S. and Alexander, R. McN. (1980) The gaits of chelonians: walking techniques for very low speeds. *J. Zool., Lond.* **191**, 353–378.

Jenkins, F. A. (1971) Limb posture and locomotion in the Virginia opossum (*Didelphis marsupialis*) and in other non-cursorial mammals. *J. Zool., Lond.* **165**, 303–315.

Ker, R. F. (1981) Dynamic tensile properties of the plantaris tendon of sheep (*Ovis aries*). *J. exp. Biol.* **93**, 283–302.

Morgan, D. L., Proske, U. and Warren, D. (1978) Measurements of muscle stiffness and the mechanism of elastic storage of energy in hopping kangaroos. *J. Physiol., Lond.* **282**, 253–261.

Rewcastle, S. C. (1980) Form and function in lacertilian knee and mesotarsal joints; a contribution to the analysis of sprawling locomotion. *J. Zool., Lond.* **191**, 147–170.

Stork, N. E. (1980) Experimental analysis of adhesion of *Chrysolina polita* (Chrysomelidae: Coleoptera) on a variety of surfaces. *J. exp. Biol.* **88**, 91–107.

Taylor, C. R. (1980) Mechanical efficiency of terrestrial locomotion: a useful concept? In Elder and Trueman (1980) (*see* General) 235–244.

Chapter 6

Allen, R. D. (1973) Biophysical aspects of pseudopodium formation and retraction, in *The Biology of Amoeba* (ed. K. W. Jeon), Academic Press, New York, 201–247.

Denny, M. (1980) The role of gastropod pedal mucus in locomotion. *Nature* **285**, 160–161.

Elder, H. Y. (1973) Direct peristaltic progression and the functional significance of the dermal connective tissue during burrowing in the polychaete *Polyphysia crassa* (Oersted). *J. exp. Biol.* **58**, 637–655.

Jones, H. D. and Trueman, E. R. (1970) Locomotion of the limpet, *Patella vulgata* L. *J. exp. Biol.* **52**, 201–216.

Gans, C. (1962) Terrestrial locomotion without limbs. *Am. Zool.* **2**, 167–182.

Seymour, M. K. (1969) Locomotion and coelomic pressure in *Lumbricus terrestris* L. *J. exp. Biol.* **51**, 47–58.

Trueman, E. R. (1967) The dynamics of burrowing in *Ensis* (Bivalvia). *Proc. R. Soc. B.* **166**, 459–476.
Trueman, E. R. (1975) *The Locomotion of Soft-Bodied Animals.* Edward Arnold, London.

Chapter 7

Baker, R. R. (1978) *The Evolutionary Ecology of Animal Migration.* Hodder & Stoughton, London.
Hamilton, W. D. (1971) Geometry for the selfish herd. *J. theoret. Biol.* **31**, 295–311.
Harestad, A. S. and Bunnell, F. L. (1979) Home range and body weight—a re-evaluation. *Ecology* **60**, 389–402.
Howland, H. C. (1974) Optimal strategies for predator avoidance: the relative importance of speed and manoeuvrability. *J. theoret. Biol.* **47**, 333–350.
Kenward, R. E. (1978) Hawks and doves: factors affecting success and selection in goshawk attacks on woodpigeons. *J. animal Ecol.* **47**, 449–460.
Lissaman, P. B. S. and Shollenberger, C. A. (1970) Formation flight of birds. *Science* **168**, 1003–1005.
Partridge, B. L. and Pitcher, T. J. (1979) Evidence against a hydrodynamic function for fish schools. *Nature* **279**, 418–419.
Schmidt-Nielsen, K. (1972) Locomotion: energy cost of swimming, flying and running. *Science* **177**, 222–228.
Urbani, C. B. (1979) Territory in social insects, in *Social Insects* (ed. H. R. Hermann) **1**, 91–120. Academic Press, New York.
Weihs, D. (1973) Hydromechanics of fish schooling. *Nature* **241**, 290–291.

Appendix

Alexander, R. McN. (1968) *Animal Mechanics.* Sidgwick & Jackson, London.
Batchelor, G. K. (1967) *An Introduction to Fluid Dynamics.* Cambridge University Press, Cambridge.
Prandtl, L. and Tietjens, O. G. (1957) *Applied Hydro- and Aerodynamics.* Dover, New York.
Prentis, J. M. (1970) *Dynamics of Mechanical Systems.* Longman, London.
Wainwright, S. A., Biggs, W. D., Currey, J. D. and Gosline, J. M. (1976). *Mechanical Design in Organisms.* Edward Arnold, London.

Index

* Italicized page numbers refer to figure legends.

159